THE
McDonaldization
OF Society

20th ANNIVERSARY EDITION

To Alan Ritzer, who helped open my eyes to McDonaldization, and to Paul O'Connell, whose gentleness and Shavian wit were inspirations for this book.

THE
McDonaldization
OF Society

20th ANNIVERSARY EDITION

GEORGE RITZER

University of Maryland

Los Angeles | London | New Delhi
Singapore | Washington DC

Los Angeles | London | New Delhi
Singapore | Washington DC

FOR INFORMATION:

SAGE Publications, Inc.
2455 Teller Road
Thousand Oaks, California 91320
E-mail: order@sagepub.com

SAGE Publications Ltd.
1 Oliver's Yard
55 City Road
London EC1Y 1SP
United Kingdom

SAGE Publications India Pvt. Ltd.
B 1/I 1 Mohan Cooperative Industrial Area
Mathura Road, New Delhi 110 044
India

SAGE Publications Asia-Pacific Pte. Ltd.
3 Church Street
#10-04 Samsung Hub
Singapore 049483

Acquisitions Editor: Dave Repetto

Editorial Assistant: Lydia Balian

Production Editor: Eric Garner

Copy Editor: Gillian Dickens

Typesetter: C&M Digitals (P) Ltd.

Proofreader: Theresa Kay

Indexer: Jean Casalegno

Cover Designer: Scott Van Atta

Marketing Manager: Jordan Bell

Permissions Editor: Karen Ehrmann

Printed in the United States of America

Library of Congress Cataloging-in-Publication Data

Ritzer, George.

The McDonaldization of society / George Ritzer.—20th anniversary edition

p. cm.
Includes bibliographical references and index.

ISBN 978-1-4522-2669-9 (pbk.)

1. Social structure—United States. 2. United States—Social conditions—1980- 3. Management—Social aspects—United States. 4. Fast food restaurants—Social aspects—United States. 5. Rationalization (Psychology) I. Title.

HM706.R58 2013
302.3'50973—dc23   2012003917

This book is printed on acid-free paper.

Certified Chain of Custody
SUSTAINABLE   Promoting Sustainable Forestry
FORESTRY   www.sfiprogram.org
INITIATIVE   SFI-01268
SFI label applies to text stock

12 13 14 15 16 10 9 8 7 6 5 4 3 2 1

Contents

Preface

This seventh edition of *The McDonaldization of Society* is being published on its 20th anniversary. In the previous revisions, I have concentrated on adding new material and bringing in new perspectives. As a result, the book has grown in length and complexity. As I thought about this new edition, my initial inclination was to do the same in this iteration. Some new text has been added, especially on the significance of In-N-Out Burger and Pret A Manger for McDonaldization and the fast-food industry, and all of the data have been brought up-to-date. However, I concluded that the most important thing to do this time around was to shorten and simplify the book without losing its basic message and main themes. As a result, the book now has far fewer pages than the previous edition. The 10 chapters of the sixth edition have been reduced to 7 chapters in this version. This has been accomplished by both cutting text and combining previously separate chapters. The dimensions of efficiency and calculability, previously treated in two separate chapters, are now combined in one concise chapter (Chapter 3), and the same is true of predictability and control (Chapter 4). Similarly, the topics of globalization and deMcDonaldization, which were dealt with in separate chapters, are now combined in a new final chapter (Chapter 7). Some significant cuts were required to accomplish this, especially in my treatment of globalization. Rather than review the literature on globalization and go into great detail on my argument in *The Globalization of Nothing*, I decided to focus on only those aspects that deal directly with the relationship between globalization and McDonaldization. In combining globalization and deMcDonaldization, I have also put together the most important issues in the continued expansion of McDonaldization (globalization) with the key challenges to that expansion. While expansion and threats to it coexist, I believe that the most likely scenario is the continued expansion of the McDonaldization of society.

1

An Introduction to McDonaldization

Ray Kroc (1902–1984), the genius behind the franchising of McDonald's restaurants, was a man with big ideas and grand ambitions. But even Kroc could not have anticipated the astounding impact of his creation. McDonald's is the basis of one of the most influential developments in contemporary society. Its reverberations extend far beyond its point of origin in the United States and in the fast-food business. It has influenced a wide range of undertakings, indeed the way of life, of a significant portion of the world. That impact is likely to continue to expand in the early 21st century.[1]*

However, this is not a book about McDonald's, or even about the fast-food business,[2] although both will be discussed frequently throughout these pages. I devote all this attention to McDonald's (as well as to the industry of which it is a part and that it played such a key role in spawning) because it serves here as the major example of, and the paradigm for, a wide-ranging process I call McDonaldization[3]—that is,

> the process by which the principles of the fast-food restaurant are coming to dominate more and more sectors of American society as well as of the rest of the world.[4]

McDonaldization has shown every sign of being an inexorable process, sweeping through seemingly impervious institutions (e.g., religion) and regions (e.g., European nations such as France) of the world.[5]

---

*Notes may be found at the back of the book, beginning on p. 186.

The success of McDonald's itself is apparent: In 2010, its revenues were $24.1 billion, with net income of $4.9 billion.[6] McDonald's, which first began operations in 1955, had almost 33,000 restaurants in 117 countries throughout the world in 2010, serving an average of 64 million customers a day.[7] A computer programmer compiled a visualization of all the McDonald's locations in America (just over 14,000 of them) and reported that it is impossible to get farther than 107 miles from a McDonald's. The "McFarthest Spot," as the programmer labeled it, lies in northwestern South Dakota.[8] A British commentator archly notes, "There are McDonald's everywhere. There's one near you, and there's one being built right now even nearer to you. Soon, if McDonald's goes on expanding at its present rate, there might even be one in your house. You could find Ronald McDonald's boots under your bed. And maybe his red wig, too."[9]

McDonald's has also been expanding its offerings with increased emphasis on its McCafé coffees, as well as with more than 1,300 McCafé locations in Europe.[10] McDonald's began to compete with Starbucks (discussed in some detail in Chapter 7) in 2007 by offering its own line of specialty coffees, and before long, a "coffee war" had been declared.

McDonald's is also employing 21st-century technologies to retain its preeminent position. More than 10,000 U.S. locations now have WiFi access. In Japan, 10 million customers receive promotional e-mails, which are "more efficient than traditional coupons." McDonald's has established an online community for its crew members, called StationM, which has blogs and other communications tools for workers to "share experiences."[11] Restaurants are being extensively remodeled and adding high-tech, drive-through lanes; big-screen televisions; video games; and even exercise bikes.

McDonald's and McDonaldization have had their most obvious influence on the restaurant industry and, more generally, on franchises of all types:

1. In 2010, small franchises had about $1.3 trillion in annual sales and accounted for more than 10% of all businesses in the United States. They employed about 8 million people.[12] Franchises are growing rapidly;[13] about 80% of McDonald's restaurants are franchises (up from 57% in 2006). (Interestingly, Starbucks refuses to franchise its operations.) In the words of McDonald's annual report, "We believe locally-owned and operated restaurants are at the core of our competitive advantage, making us not just a global brand but also a locally relevant one."[14]

2. The McDonald's model has been adopted not only by other budget-minded hamburger franchises, such as Burger King and Wendy's, but also by a wide array of other low-priced fast-food businesses. As of the beginning of 2010, Yum! Brands, Inc. operated almost 38,000 restaurants in more than 110 countries,[15] such as Pizza Hut, Kentucky Fried Chicken, Taco Bell, A&W Root Beer,

Wing Street, and Long John Silver's franchises. Yum! Brands has more outlets than McDonald's, although its total sales ($11.3 billion in 2010) and net income ($1.8 billion) are not nearly as high.[16] Subway (with more than 35,000 outlets in 98 countries)[17] is one of the fastest-growing fast-food businesses and claims to be—and may actually be—the largest restaurant chain in the United States.[18] The Cleveland, Ohio, market, to take one example, is so saturated with Subway restaurants that one opened *inside* the Jewish Community Center.[19]

3. The McDonald's model has been extended to casual dining—that is, more upscale, higher-priced chain restaurants with fuller menus (for example, Outback Steakhouse, Chili's, Olive Garden, Cheesecake Factory, and Red Lobster). Morton's is an even more upscale, high-priced chain of steak-houses that has overtly modeled itself after McDonald's: "Despite the fawning service and the huge wine list, a meal at Morton's conforms to the same dictates of uniformity, cost control and portion regulation that have enabled American fast-food chains to rule the world."[20] In fact, the chief executive of Morton's was an owner of a number of Wendy's outlets and admits, "My experience with Wendy's has helped in Morton's venues."[21] To achieve uniformity, employees go "by the book": "an ingredient-by-ingredient illustrated binder describing the exact specifications of . . . Morton's kitchen items, sauces and garnishes. A row of color pictures in every Morton's kitchen displays the presentation for each dish."[22,23] Each Morton's also offers private boardrooms with standardized features, including "state-of-the-art, high-definition satellite television broadcast reception, large drop-down screens and theater-quality surround sound, Wi-Fi technology and Velocity broadcasting capabilities." At the end of 2010, Morton's owned and operated 77 steakhouses (Morton's does not franchise) with an annual revenue of $296 million.[24]

4. Other types of business are increasingly adapting the principles of the fast-food industry to their needs. Said the vice chairman of Toys"R"Us, "We want to be thought of as a sort of McDonald's of toys."[25] (Interestingly, Toys"R"Us is now in decline because of its inability to compete with the even more McDonaldized Wal-Mart and its toy business.) The founder of Kidsports Fun and Fitness Club echoed this desire: "I want to be the McDonald's of the kids' fun and fitness business."[26] Other chains with similar ambitions include Gap, Jiffy Lube, AAMCO Transmissions, Midas Muffler & Brake Shops, Great Clips, H&R Block, Pearle Vision, Bally's, Kampgrounds of America (KOA), KinderCare (dubbed "Kentucky Fried Children"),[27] Jenny Craig, Home Depot, and PetSmart. Curves, the world's largest chain of women's fitness centers, was founded in 1995, and by 2010, there were nearly 10,000 of them in 50 states and more than 85 countries.[28] The company touts the fact that "there is approximately 1 Curves for every 2 McDonald's in the U.S."[29] The company also claims, "What took McDonalds 25 years and Subway 26 years to do—open 7,000 locations—Curves did in under a decade."[30]

5. McDonald's has been a resounding success in the international arena. More than 42% of McDonald's restaurants are outside the United States (in the mid-1980s, only 25% of McDonald's were outside the United States).[31] The majority of the new restaurants being opened are overseas.[32] Well over half of McDonald's revenue comes from its overseas operations. The leader by far, as of 2009, is Japan with 3,754 restaurants.[33] There are currently more than 2,000 McDonald's restaurants in China, and the company plans on a major expansion there.[34] (However, Yum! Brands operates more than 3,200 KFCs—the Chinese greatly prefer chicken to beef—and 500 Pizza Huts in China.[35] Furthermore, Yum! Brands is expanding faster in China than McDonald's. It has 40% of the fast-food restaurant market compared to McDonald's 16%).[36] As of 2010, there were 240 McDonald's in Russia, and it is the company's fastest growing market.[37] McDonald's plans to open many more restaurants in the former Soviet Union and in the vast new territory in Eastern Europe that has been laid bare to the invasion of fast-food restaurants. In fact, many other fast-food restaurants are succeeding in Russia; Russians seem to love American fast food.[38] Although there have been recent setbacks for McDonald's in Great Britain, that nation remains the "fast-food capital of Europe,"[39] and Israel is described as "McDonaldized," with its shopping malls populated by "Ace Hardware, Toys 'R Us, Office Depot, and TCBY."[40]

6. Many highly McDonaldized firms outside the fast-food industry have also had success globally. Wal-Mart is the world's largest retailer with 2.1 million employees (1.4 million in the United States) and almost $419 billion in sales in 2011. About 3,800 of its stores are in the United States (as of 2011). It opened its first international store (in Mexico) in 1991; it now has more than 4,557 stores worldwide, including Argentina, Brazil, Canada, Chile, China, Costa Rica, El Salvador, Guatemala, Honduras, India, Japan, Mexico, Nicaragua, Puerto Rico, and the United Kingdom.[41] In any given week, more than 175 million customers visit Wal-Mart stores worldwide.[42]

7. Other nations have developed their own variants on the McDonald's chain. Canada has a chain of coffee shops called Tim Hortons (merged with Wendy's in 1995), with 3,811 outlets (622 in the United States) in mid-2011.[43] It is Canada's largest food service provider with nearly twice as many outlets as McDonald's in that country. The chain has 62% of Canada's coffee business (Starbucks is a distant second with just 7% of that business).[44] Paris, a city whose love for fine cuisine might lead you to think it would prove immune to fast food, has a large number of fast-food croissanteries; the revered French bread has also been McDonaldized.[45] India has a chain of fast-food restaurants, Nirula's, that sells mutton burgers (about 80% of Indians are Hindus, who eat no beef) as well as local Indian cuisine.[46] Mos Burger is a Japanese chain with more than 1,600 restaurants in eight countries[47] that, in addition to the usual fare, sell Teriyaki chicken burgers, rice burgers, and "Oshiruko with brown rice cake."[48] Perhaps the most unlikely spot for an indigenous fast-food restaurant, war-ravaged Beirut of 1984, witnessed the opening of Juicy Burger, with a

rainbow instead of golden arches and J. B. the Clown standing in for Ronald McDonald. Its owners hoped it would become the "McDonald's of the Arab world."[49] In the immediate wake of the 2003 invasion of Iraq, clones of McDonald's (sporting names like "MaDonal" and "Matbax") opened in that country complete with hamburgers, French fries, and even golden arches.[50]

8. And now McDonaldization is coming full circle. Other countries with their own McDonaldized institutions have begun to export them to the United States. As of 2011, the Body Shop, an ecologically sensitive British cosmetics chain, had more than 2,500 shops in 60 nations,[51] 300 of them in the United States. American firms, such as Bath & Body Works, have followed the lead and opened copies of this British chain.[52] Pret A Manger (see below), a chain of sandwich shops that also originated in Great Britain (interestingly, McDonald's purchased a 33% minority share of the company in 2001 but divested itself in 2008), has more than 259 company-owned and -run restaurants, mostly in the United Kingdom but now also in the United States (39 restaurants); it plans on opening soon in Paris.[53] Pollo Campero was founded in Guatemala in 1971 and by 2011 had 340 restaurants in Latin America and six other countries.[54] More than 50 Pollo Campero restaurants are operating in the United States. Jollibee, a Philippine chain, has 600 stores, with 25 U.S. outlets.[55] Although Pollo Campero is a smaller presence in the United States than the American-owned Pollo Tropical chain (which has a total of 120 outlets, almost all in the United States, including Puerto Rico, as well as a smattering in Latin America),[56] Pollo Campero is more significant because it involves the invasion of the United States, the home of fast food, by a foreign chain.

9. IKEA (more on this important chain later), a Swedish-based (but Dutch-owned) home furnishings company, did about $23.1 billion of business in 2010, derived from the more than 660 million people visiting its 326 stores in 38 countries.[57] Purchases were also made from the 197 million copies of its catalog printed in 61 editions and 29 languages.[58] In fact, that catalog is reputed to print annually the second largest number of copies in the world, just after the Bible.[59] IKEA's website features more than 9,500 products and reported over 450 million visitors in 2008.[60] IKEA is so popular in Europe that "it is said that one in ten Europeans is conceived on an IKEA bed."[61] Another international chain to watch in the coming years is H&M clothing, which was founded in 1947 and now has 2,200 shops in 41 countries.[62] It currently employs more than 50,000 people and did about $12.7 billion in sales in 2007.[63] Zara opened its first shop in 1975 and now has 1,540 shops in 78 countries. Zara is part of a large group of fashion retailers that make up the Spanish firm Inditex. Taken together, there are more than 5,000 Inditex shops in the world operating under eight different brand names.[64] Inditex employed more than 89,000 people in 2008 and sold over 697 million garments with a total sales figure of more than 10.4 billion euros.[65]

10. Much of the above emphasizes the spatial expansion of McDonald's and other McDonaldized businesses, but in addition they have all expanded temporally.

McDonald's has shifted some of its attention from adding locations to adding hours to existing locales, thereby squeezing greater profits from each of them. For example, McDonald's did not at first offer breakfast, but now that meal has become the most important part of the day, and McDonald's dominates the fast-food breakfast market (although Starbucks sought, highly unsuccessfully, to challenge its preeminence). There is also a trend toward remaining open on a 24/7 basis. While fewer than 1% of McDonald's restaurants in the United States operated nonstop in 2002, almost 40% were operating that way by 2009. Moreover, 80% of U.S. locations now open by 5 a.m.[66] Time, like space, is no barrier to the spread of McDonald's and McDonaldization.

11. As we will see throughout this book, it is possible to view a wide range of the most contemporary phenomena as being affected directly or indirectly by the McDonald's model (and McDonaldization). Among them are text messaging, multitasking, iPhones, iPods, Facebook, YouTube, eBay, Craigslist, Second Life, online dating (e.g., match.com), Viagra, virtual vacations, and extreme sports.

McDonald's as an American and a Global Icon

McDonald's has come to occupy a central place not just in the business world but also in American and global popular culture.[67] The opening of a new McDonald's in a small town can be an important social event. Said one Maryland high school student at such an opening, "Nothing this exciting ever happens in Dale City."[68] Even big-city and national newspapers avidly cover developments in the fast-food business.

Fast-food restaurants also play symbolic roles on television programs and in the movies. A skit on the legendary television show *Saturday Night Live* satirized specialty chains by detailing the hardships of a franchise that sold nothing but Scotch tape. In the movie *Coming to America* (1988), Eddie Murphy plays an African prince whose introduction to America includes a job at "McDowell's," a thinly disguised McDonald's. In *Falling Down* (1993), Michael Douglas vents his rage against the modern world in a fast-food restaurant dominated by mindless rules designed to frustrate customers. In *Sleeper* (1973), Woody Allen awakens in the future only to encounter a McDonald's. *Tin Men* (1987) ends with the early 1960s heroes driving off into a future represented by a huge golden arch looming in the distance. *Scotland, PA* (2001) brings Shakespeare's *Macbeth* to the Pennsylvania of the 1970s. The famous murder scene involves, in this case, plunging a doughnut king's head into the boiling oil of a deep fat fryer. The "McBeths" then use their ill-gotten gains to transform the king's greasy spoon café into a fast-food restaurant featuring "McBeth" burgers. The focus of the movie *Fast Food Nation* (2006) is a fictional fast-food chain ("Mickey's") and its hit hamburger ("The Big One"), the beef processor that supplies the meat, and the plight of the illegal Mexican immigrants who work there. In the 2008 remake of the sci-fi classic *The Day the Earth Stood Still*, an important meeting, and perhaps the pivotal scene in the movie, takes place in McDonald's. There, a newly arrived alien meets another who has been on Earth for decades and is dissuaded from destroying humanity by the fact that the latter has learned to love humans.

When plans were made to raze Ray Kroc's first McDonald's restaurant, hundreds of letters poured into company headquarters, including the following: "Please don't tear it down! . . . To destroy this major artifact of contemporary culture would, indeed, destroy part of the faith the people of the world have in your company."[69] In the end, the restaurant was rebuilt according to the original blueprints and turned into a museum.[70] A McDonald's executive explained the move: "McDonald's . . . is really a part of Americana."

Americans aren't the only ones who feel this way. At the opening of the McDonald's in Moscow, one journalist described the franchise as the "ultimate icon of Americana."[71] When Pizza Hut opened in Moscow, a Russian student said, "It's a piece of America."[72] Reflecting on the growth of fast-food restaurants in Brazil, an executive associated with Pizza Hut of Brazil said that his nation "is experiencing a passion for things American."[73] On the popularity of Kentucky Fried Chicken in Malaysia, the local owner said, "Anything Western, especially American, people here love. . . . They want to be associated with America."[74]

One could go further and argue that in at least some ways, McDonald's has become more important than the United States itself. Take the following story about a former U.S. ambassador to Israel officiating at the opening of the first McDonald's in Jerusalem wearing a baseball cap with the McDonald's golden arches logo:

> An Israeli teen-ager walked up to him, carrying his own McDonald's hat, which he handed to Ambassador Indyk with a pen and asked: "Are you the Ambassador? Can I have your autograph?" Somewhat sheepishly, Ambassador Indyk replied: "Sure. I've never been asked for my autograph before."
>
> As the Ambassador prepared to sign his name, the Israeli teen-ager said to him, "Wow, what's it like to be the ambassador from McDonald's, going around the world opening McDonald's restaurants everywhere?"
>
> Ambassador Indyk looked at the Israeli youth and said, "No, no. I'm the American ambassador—not the ambassador from McDonald's!" Ambassador Indyk described what happened next: "I said to him, 'Does this mean you don't want my autograph?' And the kid said, 'No, I don't want your autograph,' and he took his hat back and walked away."[75]

Two other indices of the significance of McDonald's (and, implicitly, McDonaldization) are worth mentioning. The first is the annual "Big Mac

Index" (part of "burgernomics"), published, tongue-in-cheek, by a prestigious magazine, *The Economist*. It indicates the purchasing power of various currencies around the world based on the local price (in dollars) of the Big Mac. The Big Mac is used because it is a uniform commodity sold in many different nations. In the 2011 survey, a Big Mac in the United States cost an average of $4.07; in China, it was $2.27; and in Switzerland, it cost $8.06.[76] This measure indicates, at least roughly, where the cost of living is high or low, as well as which currencies are undervalued (China) and overvalued (Switzerland). Although *The Economist* is calculating the Big Mac Index only half-seriously, the index represents the ubiquity and importance of McDonald's around the world.[77] Alternatively, *The Economist* measured economic disparity by comparing the labor time required for the average workers in various cities to earn enough to purchase a Big Mac. The least amount of labor time—12 minutes— was required in Chicago, while workers in Nairobi had to work for nearly 160 minutes to be able to buy a Big Mac.[78]

The second indicator of the global significance of McDonald's is the idea developed by Thomas Friedman that "no two countries that both have a McDonald's have ever fought a war since they each got McDonald's." Friedman calls this the "Golden Arches Theory of Conflict Prevention."[79] Another tongue-in-cheek idea, it implies that the path to world peace lies through the continued international expansion of McDonald's. Unfortunately, it was proved wrong by the NATO bombing of Serbia in 1999, which had McDonald's at the time.

To many people throughout the world, McDonald's has become a sacred institution.[80] At that opening of the McDonald's in Moscow, a worker spoke of it "as if it were the Cathedral in Chartres, . . . a place to experience 'celestial joy.'"[81] Kowinski argues that indoor shopping malls, which almost always encompass fast-food restaurants and other franchises and chains, are the modern "cathedrals of consumption" to which people go to practice their "consumer religion."[82] Similarly, a visit to another central element of McDonaldized society, Walt Disney World,[83] has been described as "the middle-class hajj, the compulsory visit to the sunbaked holy city."[84]

McDonald's has achieved its exalted position because virtually all Americans, and many others, have passed through its golden arches (or by its drive-through windows) on innumerable occasions. Furthermore, most of us have been bombarded by commercials extolling the virtues of McDonald's, commercials tailored to a variety of audiences and that change as the chain introduces new foods, new contests, and new product tie-ins. These ever-present commercials, combined with the fact that people cannot drive or walk very far without having a McDonald's pop into view, have embedded McDonald's deeply in popular consciousness. A poll of school-age children showed that 96% of them could identify Ronald McDonald, second only to Santa Claus in name recognition.[85]

Over the years, McDonald's has appealed to people in many ways. The restaurants themselves are depicted as spick-and-span, the food is said to be fresh and nutritious, the employees are shown to be young and eager, the managers appear gentle and caring, and the dining experience itself seems fun-filled. Through their purchases, people contribute, at least indirectly, to charities such as the Ronald McDonald Houses for sick children.

The Long Arm of McDonaldization

McDonald's strives continually to extend its reach within American society and beyond. As the company's chairman said, "Our goal: to totally dominate the quick service restaurant industry worldwide. . . . I want McDonald's to dominate."[86]

McDonald's began as a phenomenon of suburbs and medium-sized towns, but later it moved into smaller towns that supposedly could not support such a restaurant and into many big cities that were supposedly too sophisticated.[87] Today, you can find fast-food outlets in New York's Times Square. McDonald's can even be found on the Guantanamo Bay U.S. Naval Base in Cuba and in the Pentagon. Small, satellite, express, or remote outlets, opened in areas that could not support full-scale fast-food restaurants, are also expanding rapidly. They are found in small storefronts in large cities and in nontraditional settings such as museums, department stores, service stations,[88] and even schools. These satellites typically offer only limited menus and may rely on larger outlets for food storage and preparation.[89] A flap arose over the placement of a McDonald's in the new federal courthouse in Boston.[90]

In Paris, McDonald's is not only on the Champs-Elysées,[91] but there is even a branch *in* the Louvre. Soon after it opened in 1992, the McDonald's in Moscow's Pushkin Square sold almost 30,000 hamburgers a day and employed a staff of 1,200 young people working two to a cash register.[92,93] In early 1992, Beijing witnessed the opening of what still may be the world's largest McDonald's, with 700 seats, 29 cash registers, and nearly 1,000 employees. On its first day of business, it set a new one-day record for McDonald's by serving about 40,000 customers.[94] Among the more striking sites for a McDonald's restaurant are at the Grand Canyon; in what was at the time the world's tallest building, the Petronas Towers in Malaysia; as a ski-through on a slope in Sweden; and in a structure in Shrewsbury, England, that dates back to the 13th century.

No longer content to dominate the strips that surround many college campuses, fast-food restaurants have moved right onto many of those campuses. The first campus fast-food restaurant opened at the University of Cincinnati in 1973. Today, college cafeterias often look like shopping mall food courts (and it's no wonder, given that campus food service is a multi-billion-dollar-a-year business).[95] In conjunction with a variety of "branded partners" (for example, Pizza Hut and Subway), Marriott now supplies food to many colleges and universities.[96] The apparent approval of college administrations puts fast-food restaurants in a position to further influence the younger generation.

We no longer need to leave many highways to obtain fast food quickly and easily. Fast food is now available at many convenient rest stops along the road. After "refueling," we can proceed with our trip, which is likely to end in another community with about the same density and mix of fast-food restaurants as the locale we left behind. Fast food is also increasingly available in hotels,[97] railway stations, and airports.

In other sectors of society, the influence of fast-food restaurants has been subtler but no less profound. Food produced by McDonald's and other fast-food restaurants has begun to appear in high schools and trade schools; more than 50% of school cafeterias offer popular brand-name fast foods such as McDonald's, Pizza Hut, or Taco Bell at least once a week.[98] Said the director of nutrition for the American School Food Service Association, "Kids today live in a world where fast food has become a way of life. For us to get kids to eat, period, we have to provide some familiar items."[99] Few lower-grade schools as yet have in-house fast-food restaurants; however, many have had to alter school cafeteria menus and procedures to make fast food readily available.[100] Apples, yogurt, and milk may go straight into the trash can, but hamburgers, fries, and shakes are devoured. Fast-food restaurants also tend to cluster within walking distances of schools.[101] The attempt to hook school-age children on fast food reached something of a peak in Illinois, where McDonald's operated a program called "A for Cheeseburger." Students who received As on their report cards received a free cheeseburger, thereby linking success in school with McDonald's.[102] In Australia, toy versions of food featured by McDonald's have been marketed to children as young as 3. The toys include "fake McDonald's fries, a self-assembling Big Mac, milkshake, Chicken McNuggets, baked apple pie and mini cookies."[103] Many fear that playing with such toy food will increase children's interest in eating the real thing.

The military has also been pressed to offer fast food on both bases and ships. Despite criticisms by physicians and nutritionists, fast-food outlets have turned up inside U.S. general hospitals and in children's hospitals.[104] While no private homes yet have a McDonald's of their own, meals at home often resemble those available in fast-food restaurants. Frozen, microwavable, and prepared foods, which bear a striking resemblance to meals available at fast-food restaurants, often find their way to the dinner table. There are even cookbooks—for example, *Secret Fast Food Recipes: The Fast Food Cookbook*—that allow one to prepare "genuine" fast food at home.[105] Then there is also home delivery of fast foods, especially pizza, as revolutionized by Domino's.

Another type of expansion involves what could be termed "vertical McDonaldization";[106] that is, the demands of the fast-food industry, as is well documented in Eric Schlosser's *Fast Food Nation,* have forced industries that service it to McDonaldize in order to satisfy its insatiable demands. Potato growing and processing, cattle ranching, chicken raising, and meat slaughtering and processing have all had to McDonaldize their operations, leading to dramatic increases in production. That growth has not come without costs, however.

As demonstrated in the movie *Food, Inc.* (2009), meat and poultry are more likely to be disease ridden, small (often non-McDonaldized) producers and ranchers have been driven out of business, and millions of people have been forced to work in low-paying, demeaning, demanding, and sometimes outright dangerous jobs. For example, in the meatpacking industry, relatively safe, unionized, secure, manageable, and relatively high-paying jobs in firms with once-household names such as Swift and Armour have been replaced with unsafe, nonunionized, insecure, unmanageable, and relatively low-paying positions with largely anonymous corporations. While some (largely owners, managers, and stockholders) have profited enormously from vertical McDonaldization, far more have been forced into a marginal economic existence.

McDonald's is such a powerful model that many businesses have acquired nicknames beginning with "Mc." Examples include "McDentists" and "McDoctors," meaning drive-in clinics designed to deal quickly and efficiently with minor dental and medical problems;[107] "McChild" care centers, meaning childcare centers such as KinderCare; "McStables," designating the nationwide racehorse-training operation of D. Wayne Lucas; and "McPaper," describing the newspaper *USA TODAY.*[108]

McDonald's is not always enamored of this proliferation. Take the case of We Be Sushi, a San Francisco chain with a half-dozen outlets. A note appears on the back of the menu explaining why the chain was not named "McSushi":

> The original name was McSushi. Our sign was up and we were ready to go. But before we could open our doors we received a very formal letter from the lawyers of, you guessed it, McDonald's. It seems that McDonald's has cornered the market on every McFood name possible from McBagle [sic] to McTaco. They explained that the use of the name McSushi would dilute the image of McDonald's.[109]

So powerful is McDonaldization that the derivatives of McDonald's, in turn, exert their own powerful influence. For example, the success of *USA TODAY* led many newspapers across the nation to adopt shorter stories and colorful weather maps. As one *USA TODAY* editor said, "The same newspaper editors who call us McPaper have been stealing our McNuggets."[110]

Even serious journalistic enterprises such as the *New York Times* and *Washington Post* have undergone changes (for example, the use of color) as a result of the success of *USA TODAY*. The influence of *USA TODAY* is blatantly manifested in many local newspapers throughout the United States.[111] As in *USA TODAY*, stories usually start and finish on the same page. Many important details, much of a story's context, and much of what the principals have to say are cut back severely or omitted entirely. With its emphasis on light news and color graphics, the main function of such a newspaper seems to be entertainment.

Like virtually every other sector of society, sex has been McDonaldized.[112] In New York City, an official called a three-story pornographic center "the McDonald's of sex" because of its "cookie-cutter cleanliness and compliance with the law."[113] In the movie *Sleeper*, Woody Allen not only created a futuristic world in which McDonald's was an important and highly visible element, but he also envisioned a society in which people could enter a machine called an "orgasmatron" to experience an orgasm without going through the muss and fuss of sexual intercourse.

The porn site RedTube mimics the standardized interface of YouTube to provide various categories of adult content that users can view on the site or embed in their own Web pages. The Web is filled with video chat sites where users can request the performance of various sex acts. The casual encounters section on Craigslist.org provides people from every city in the world with a centralized interface to find sexual partners. A variety of devices, termed "teledildonics" by the adult entertainment industry, enable users to stimulate one another through computer networks. 3Feel is a virtual 3D environment where users can interact in real time and engage in sexual activity (with or without teledildonics).[114] As Woody Allen anticipated with his orgasmatron, "Participants can experience an orgasm without ever meeting or touching one another."[115]

> In a world where convenience is king, disembodied sex has its allure. You don't have to stir from your comfortable home. You pick up the phone, or log onto the computer and, if you're plugged in, a world of unheard of sexual splendor rolls out before your eyes.[116]

These examples suggest that no aspect of people's lives is immune to McDonaldization.

Various pharmaceuticals can be seen as McDonaldizing sex. Viagra (and similar drugs such as Cialis) do this by, for example, making the ability of males to have sex more predictable. Such drugs also claim to work fast and to last for a long time. MDMA (ecstasy) lasts for as much as 8 hours and tends to increase the intensity of sensory information and feelings of social (including sexual) connectedness.

The preceding merely represents the tip of the iceberg as far as the long arm of McDonaldization is concerned. Other areas affected by McDonaldization include the following:[117]

- Mountain climbing (e.g., reliance on guidebooks to climbing routes)[118]
- Criminal justice system (police profiling, "three strikes and you're out")[119]
- Family (books, TV shows devoted to quick fixes to family problems)[120]
- McSchools and the policies that serve to McDonaldize them[121]
- Losing weight and the McDonaldization of the body[122]
- Internet as a site of McDonaldization (and deMcDonaldization—see Chapter 7)[123]
- Farms and their supersizing[124]
- Religion and the McDonaldization of religious creeds[125]
- McJobs[126]
- Politics ("cool" vs. "hot" politics; "drive-through democracy")[127]

The Dimensions of McDonaldization

Why has the McDonald's model proven so irresistible? Eating fast food at McDonald's has certainly become a "sign"[128] that, among other things, one is in tune with the contemporary lifestyle. There is also a kind of magic or enchantment associated with such food and its settings. The focus here, however, is on the four dimensions that lie at the heart of the success of this model and, more generally, of McDonaldization. In short, McDonald's has succeeded because it offers consumers, workers, and managers efficiency, calculability, predictability, and control.[129] Chapters 3 and 4 will be devoted to these dimensions of McDonaldization, but it is important to at least mention them at this point.

Efficiency

One important element of the success of McDonald's is *efficiency*, or the optimum method for getting from one point to another. For consumers, McDonald's (its drive-through is a good example) offers the best available way to get from being hungry to being full. The fast-food model offers, or at least appears to offer, an efficient method for satisfying many other needs, as well. Woody Allen's orgasmatron offered an efficient method for getting from quiescence to sexual gratification. Other institutions fashioned on the McDonald's model offer similar efficiency in exercising, losing weight, lubricating cars, getting new glasses or contacts, or completing income tax forms. Like their customers, workers in McDonaldized systems function efficiently by following the steps in a predesigned process.

Calculability

Calculability emphasizes the quantitative aspects of products sold (portion size, cost) and services offered (the time it takes to get the product). In McDonaldized systems, quantity has become equivalent to quality; a lot of something, or the quick delivery of it, means it must be good. "As a culture, we tend to believe deeply that in general 'bigger is better.'"[130] People can quantify things and feel that they are getting a lot of food for what appears to be a nominal sum of money (best exemplified by the McDonald's "Dollar Menu").[131] In a Denny's ad, a man says, "I'm going to eat too much, but I'm never going to pay too much."[132] This calculation does not take into account an important point, however: The high profit margin of fast-food chains indicates that the owners, not the consumers, get the best deal.

People also calculate how much time it will take to drive to McDonald's, be served the food, eat it, and return home; they then compare that interval to the time required to prepare food at home. They often conclude, rightly or wrongly, that a trip to the fast-food restaurant will take less time than eating at home. This sort of calculation particularly supports home delivery franchises such as Domino's, as well as other chains that emphasize saving time. A notable example of time savings in another sort of chain is LensCrafters, which promises people "Glasses fast, glasses in one hour." H&M is known for its "fast fashion."

Some McDonaldized institutions combine the emphases on time and money. Domino's promises pizza delivery in half an hour, or the pizza is free. Pizza Hut will serve a personal pan pizza in 5 minutes, or it, too, will be free.

Workers in McDonaldized systems also emphasize the quantitative rather than the qualitative aspects of their work. Since the quality of the work is allowed to vary little, workers focus on how quickly tasks can be accomplished. In a situation analogous to that of the customer, workers are expected to do a lot of work, very quickly, for low pay.

Predictability

McDonald's also offers *predictability*, the assurance that products and services will be the same over time and in all locales. Egg McMuffins in New York will be virtually identical to those in Chicago and Los Angeles. Also, those eaten next week or next year will be about the same as those eaten today. Customers take great comfort in knowing that McDonald's offers no surprises. They know that the next Egg McMuffin they eat will not be awful, but it will not be exceptionally delicious, either. The success of the McDonald's model suggests that many people have come to prefer a world in which there are few surprises. "This is strange," notes a British observer, "considering [McDonald's is] the product of a culture which honours individualism above all."[133]

The workers in McDonaldized systems also behave in predictable ways. They follow corporate rules as well as the dictates of their managers. In many cases, what they do, and even what they say, is highly predictable.

Control

The fourth element in the success of McDonald's, *control*,[134] is exerted over the people who enter McDonald's. Lines, limited menus, few options, and uncomfortable seats all lead diners to do what management wishes them to do—eat quickly and leave. Furthermore, the drive-through window invites diners to leave before they eat. In the Domino's model, customers never enter in the first place.

The people who work in McDonaldized organizations are also controlled to a high degree, usually more blatantly and directly than customers. They are trained to do a limited number of tasks in precisely the way they are told to do them. This control is reinforced by the technologies used and the way the organization is set up to bolster this control. Managers and inspectors make sure that workers toe the line.

A Critique of McDonaldization: The Irrationality of Rationality

McDonaldization offers powerful advantages. In fact, efficiency, predictability, calculability, and control through nonhuman technology (that is, technology that controls people rather than being controlled by them) can be thought of not only as the basic components of a rational system[135] but also as the powerful advantages of such a system. However, rational systems inevitably spawn irrationalities. The downside of McDonaldization will be dealt with most systematically under the heading of the irrationality of rationality; in fact, paradoxically, the irrationality of rationality can be thought of as the fifth dimension of McDonaldization (see Chapter 5).

Criticism, in fact, can be applied to all facets of the McDonaldizing world. As just one example, at the opening of Euro Disney, a French politician said that it will "bombard France with uprooted creations that are to culture what fast food is to gastronomy."[136] McDonald's and other purveyors of the fast-food model spend billions of dollars each year detailing the benefits of their system. Critics of the system, however, have few outlets for their ideas. For example, no one sponsors commercials between Saturday morning cartoons warning children of the dangers associated with fast-food restaurants.

Nonetheless, a legitimate question may be raised about this critique of McDonaldization: Is it animated by a romanticization of the past, an impossible desire to return to a world that no longer exists? Some critics do base their

critiques on nostalgia for a time when life was slower and offered more surprises, when at least some people (those who were better off economically)
were freer, and when one was more likely to deal with a human being than a
robot or a computer.[137] Although they have a point, these critics have undoubtedly exaggerated the positive aspects of a world without McDonald's, and they
have certainly tended to forget the liabilities associated with earlier eras. As an
example of the latter, take the following anecdote about a visit to a pizzeria in
Havana, Cuba, which in some respects is decades behind the United States:

> The pizza's not much to rave about—they scrimp on tomato sauce, and the dough
> is mushy.
>
> It was about 7:30 P.M., and as usual the place was standing-room-only, with
> people two deep jostling for a stool to come open and a waiting line spilling out
> onto the sidewalk.
>
> The menu is similarly Spartan. . . . To drink, there is tap water. That's it—no top-
> pings, no soda, no beer, no coffee, no salt, no pepper. And no special orders.
>
> A very few people are eating. Most are waiting. . . . Fingers are drumming, flies
> are buzzing, the clock is ticking. The waiter wears a watch around his belt loop,
> but he hardly needs it; time is evidently not his chief concern. After a while, tem-
> pers begin to fray.
>
> But right now, it's 8:45 P.M. at the pizzeria, I've been waiting an hour and a quar-
> ter for two small pies.[138]

Few would prefer such a restaurant to the fast, friendly, diverse offerings of,
say, Pizza Hut. More important, however, critics who revere the past do not
seem to realize that we are not returning to such a world. In fact, fast-food
restaurants have begun to appear even in Havana (and many more are likely
after the death of Fidel Castro).[139] The increase in the number of people crowd-
ing the planet, the acceleration of technological change, the increasing pace of
life—all this and more make it impossible to go back to the world, if it ever
existed, of home-cooked meals, traditional restaurant dinners, high-quality
foods, meals loaded with surprises, and restaurants run by chefs free to express
their creativity.

   It is more valid to critique McDonaldization from the perspective of a con-
ceivable future.[140] Unfettered by the constraints of McDonaldized systems, but
using the technological advances made possible by them, people could have the
potential to be far more thoughtful, skillful, creative, and well rounded than
they are now. In short, if the world were less McDonaldized, people would be
better able to live up to their human potential.

   We must look at McDonaldization as both "enabling" and "constrain-
ing."[141] McDonaldized systems enable us to do many things we were not able

to do in the past; however, these systems also keep us from doing things we otherwise would do. McDonaldization is a "double-edged" phenomenon.

Illustrating the Dimensions of McDonaldization: The Case of IKEA

An interesting example of McDonaldization, especially since it has its roots in Sweden rather than the United States, is IKEA.[142] In fact, IKEA is so important that Tod Hartman has written about the "IKEAization of France,"[143] although it could be that it is more the IKEAization of society as a whole that is being discussed. IKEA's popularity stems from the fact that it offers, at very low prices, trendy furniture based on well-known Swedish designs. It has a large and devoted clientele throughout the world. What is interesting about IKEA from the point of view of this book is how well it fits the dimensions of McDonaldization. The similarities go beyond that, however. For example, just as with the opening of a new McDonald's, there is great anticipation over the opening of the first IKEA in a particular location. Just the rumor that one was to open in Dayton, Ohio, led to the following statement: "We here in Dayton are peeing our collective pants waiting for the IKEA announcement."[144] IKEA is also a global phenomenon that sells in many countries both its signature products as well as a few (e.g., chopsticks in China) more adapted to local tastes and interests.[145]

In terms of *efficiency*, IKEA offers one-stop furniture shopping with an extraordinary range of furniture. In general, there is no waiting for one's purchases since a huge warehouse is attached to each store (one often enters through the warehouse), with large numbers of virtually everything in stock. Much of the efficiency at IKEA stems from the fact that customers are expected to do a lot of the work:

- Unlike McDonald's, there are relatively few IKEAs in any given area; thus, customers most often spend many hours driving great distances to get to a store. This is known as the "IKEA road trip."[146]
- On entry, customers are expected to take a map to guide themselves through the huge and purposely maze-like store (IKEA hopes, like Las Vegas casinos, that customers will get "lost" in the maze and wander for hours, spending money as they go). There are few employees to guide anyone, but there are paths painted on the floor that customers can follow on their own.
- Also upon entry, customers are expected to grab a pencil and an order form and to write down the shelf and bin numbers for the larger items they wish to purchase; a yellow shopping bag is to be picked up on entry for smaller items. There is little in the way of help available as customers wander through the stores. Customers can switch from a shopping bag to a shopping cart after leaving the

showroom and entering the marketplace, where they can pick up other smaller items.

- If customers eat in the cafeteria, they are expected to clean their tables after eating. There is even this helpful sign: "Why should I clean my own table? At IKEA, cleaning your own table at the end of your meal is one of the reasons you paid less at the start."[147]
- Most of the furniture sold is unassembled in flat packages, and customers are expected to load most of the items (except the largest) into their cars themselves. After they get home, they must break down (and dispose of) the packaging and then put their furniture together. If the furniture does not fit into your car, you can rent a truck on site to transport it home or have it delivered, although the cost tends to be high, especially relative to the price paid for the furniture.
- To get a catalog,[148] customers often sign up online.

In these ways, and others, IKEA has its customers do unpaid work.

Calculability is at the heart of IKEA, especially the idea that what is offered is at a very low price: "IKEA uses a technique called 'bulla bulla,' in which a bunch of items are purposely jumbled in bins, to create the impression of volume and, therefore, inexpensiveness."[149] Like a McDonald's "Dollar Menu," one can get a lot of furniture—a roomful, even a houseful—at bargain prices. As with value meals, IKEA customers feel they are getting value for their money. (The large cafeteria offers low-priced food, including the chain's signature Swedish meatballs and 99-cent breakfasts.) However, as is always the case in McDonaldized settings, low price generally means that the quality is inferior, and it is often the case that IKEA products fall apart in relatively short order. IKEA also emphasizes the huge size of its stores, which often approach 300,000 square feet or about four to five football fields. This mammoth size leads the consumer to believe that there will be a lot of furniture offered (and there is) and that, given the store's reputation, most of it will be highly affordable.

Of course, there is great *predictability* about any given IKEA—large parking lots; a supervised children's play area (where IKEA provides personnel, because of liability issues as well as the fact that supervised children give parents more time and peace of mind to shop and spend); the masses of inexpensive, Swedish-designed furniture; exit through the warehouse and the checkout counters; boxes to take home with furniture requiring assembly; and so on.

An IKEA is a highly *controlled* environment, mainly in the sense that the maze-like structure of the store virtually forces the consumer to traverse the entire place and to see virtually everything it has to offer. If one tries to take a path other than that set by IKEA, one is likely to become lost and disoriented. There seems to be no way out that does not lead to the checkout counter, where you pay for your purchases.

There are a variety of *irrationalities* associated with the rationality of IKEA, most notably the poor quality of most of its products. Although the furniture is purportedly easy to assemble, many are more likely to think of it as "impossible-to-assemble."[150] Then there are the often long hours required to get to an IKEA, to wander through it, to drive back home, and then to assemble the purchases.

The Advantages of McDonaldization

This discussion of the fundamental characteristics of McDonaldization makes it clear that, despite irrationalities, McDonald's (and other McDonaldized systems such as IKEA) has succeeded so phenomenally for good, solid reasons. Many knowledgeable people, such as the economic columnist Robert Samuelson, strongly support the McDonald's business model. Samuelson confesses to "openly worship[ing] McDonald's," and he thinks of it as "the greatest restaurant chain in history."[151] In addition, McDonald's offers many praiseworthy programs that benefit society, such as its Ronald McDonald Houses, which permit parents to stay with children undergoing treatment for serious medical problems; job-training programs for teenagers; programs to help keep its employees in school; efforts to hire and train the disabled; the McMasters program, aimed at hiring senior citizens; an enviable record of hiring and promoting minorities; and a social responsibility program with goals of improving the environment and animal welfare.[152]

The process of McDonaldization also moved ahead dramatically undoubtedly because it has led to positive changes.[153] Here are a few specific examples of such changes:

- A wider range of goods and services is available to a much larger portion of the population than ever before.
- Availability of goods and services depends far less than before on time or geographic location; people can now do things that were impossible previously, such as text message, e-mail, arrange dates online, make online purchases, and participate in online social networks, in the middle of the night.
- People are able to acquire what they want or need almost instantaneously and get it far more conveniently.
- Goods and services are of far more uniform quality; at least some people even get better-quality goods and services than before McDonaldization.
- Far more economical alternatives to high-priced, customized goods and services are widely available; therefore, people can afford things (e.g., IKEA furniture rather than handmade furniture) they could not previously afford.
- Fast, efficient goods and services are available to a population that is working longer hours and has fewer hours to spare.
- In a rapidly changing, unfamiliar, and seemingly hostile world, the comparatively stable, familiar, and safe environment of a McDonaldized system offers comfort.

- Because of quantification, consumers can more easily compare competing products.
- Certain products (for example, exercise and diet programs) are safer in a carefully regulated and controlled system.
- People are more likely to be treated similarly, no matter what their race, sex, sexual orientation, or social class.
- Organizational and technological innovations are more quickly and easily diffused through networks of identical operators.
- The most popular products of one society are more easily disseminated to others.

What Isn't McDonaldized?

This chapter should give you a sense of McDonaldization and of the range of phenomena to be discussed throughout this book. In fact, such a wide range of phenomena can be linked to McDonaldization that you may begin to wonder what isn't McDonaldized. Is McDonaldization the equivalent of modernity? Is everything contemporary McDonaldized?

Although much of the world has been McDonaldized, at least three aspects of contemporary society have largely escaped the process:

- Those aspects traceable to an earlier, "premodern" age. A good example is the mom-and-pop grocery store.
- New businesses that have sprung up or expanded, at least in part, as a reaction against McDonaldization. For instance, people fed up with McDonaldized motel rooms in Holiday Inns or Motel 6s can stay instead in a bed-and-breakfast (see Chapter 7), which offers a room in a private home with personalized attention and a homemade breakfast from the proprietor.
- Those aspects suggesting a move toward a new, "postmodern" age. For example, in a postmodern society, "modern" high-rise housing projects make way for smaller, more livable communities.

Thus, although McDonaldization is ubiquitous, there is more to the contemporary world than McDonaldization. It is a very important social process, but it is far from the only process transforming contemporary society.

Are In-N-Out Burger and Pret A Manger Antitheses of McDonaldization?

There are degrees of McDonaldization. *McDonaldization is not an all-or-nothing process.*[154] Not all fast-food restaurants are as McDonaldized as McDonald's. In fact, fast food is not the problem. Most societies throughout history have had fast foods of all types, and they have not been problematic. There are many fine fast-food restaurants in the United States and the world (although few are part of large chains). Then there are the best of the street vendors in large cities such as New York who attract long lines of customers because of their fine and fast food and low prices. In fact, there is now an award in New York City, called the "Vendy," that goes to the best purveyors of the city's street food.[155]

In this section, however, we focus on two relatively small fast-food chains—In-N-Out Burger and Pret A Manger—that are far less McDonaldized than the highly McDonaldized chains dealt with thus far in this book. They have many devotees who love them, often for their non-McDonaldized qualities. Does this mean that these fast-food chains are the antitheses of McDonaldized systems?

In-N-Out Burger

In-N-Out Burger is a relatively small West Coast chain with roughly 250 restaurants. Like McDonald's and other fast chains, it has its roots in Southern California in the first half of the 20th century. (In-N-Out Burger began operations in 1948.) Unlike McDonald's, as well as the other major chains of fast-food restaurants, In-N-Out has not gone public; it remains under private ownership and has resisted many efforts to buy it and transform it into a public corporation. While it has expanded gradually over the years, it has, at least thus far, resisted the pressures and temptations to expand throughout the United States and to become a global operation. For this reason and many others, Stacy Perman argues that In-N-Out Burger is "the antithesis of McDonald's."[156] Similarly, the subtitle of Perman's book contends that In-N-Out is "the fast-food chain that breaks all the rules." It could be argued that those rules are the principles of McDonaldization outlined in this book and because In-N-Out Burger, at least in Perman's view, does not adhere to them, it is the antithesis of McDonald's. The objective in this section is to seek to determine whether or not this is the case.

Like all McDonaldized systems, In-N-Out Burger has certainly emphasized efficiency in various ways. Most notably, it was originally only a drive-through restaurant that focused on allowing customers to get their food as quickly as possible without ever leaving their cars. In fact, Perman points out that In-N-Out Burger billed itself as "California's first drive-through," but she goes further to contend that "in all probability it was the country's first as well."[157] The name itself—In-N-Out—implied efficiency for the customers; they could get their food and leave without the inefficiency and inconvenience of leaving their cars. Furthermore, In-N-Out Burger's early motto was "NO DELAY."[158]

In-N-Out Burger also maximized efficiency by offering a highly limited menu—"three burger items, french fries, soft drinks, lemonade and milkshakes"—an even more limited menu than McDonald's. The "public" In-N-Out Burger menu has changed little over the years, although a variety of soft drinks are now available. This highly limited menu makes In-N-Out Burger *more efficient* for both customers (less to choose from) and employees (fewer menu items to prepare) than McDonald's and its other major competitors. In fact, one of the company slogans is "Ordering as easy as 1, 2, 3." Interestingly, as we will see below, that efficiency is greatly limited by the existence of a so-called secret menu that greatly expands menu options and therefore reduces efficiency. In fact, In-N-Out Burger now calls it its "not-so-secret menu" and lists the alternatives publicly, including on its website. This serves to reduce efficiency at In-N-Out Burger, but as we will discuss below, it has served other functions that have greatly advantaged this burger chain.

In-N-Out Burger is also highly predictable in various ways. For example, one customer said, "In-N-Out always tastes the same."[159] Although the more recent restaurants offer indoor seating, the drive-throughs serve to make customer behavior highly predictable because they are not able to enter the restaurant and thereby unable to do any number of unanticipated things. The drive-through window also limited what employees could say and do, at least relative to customers. However, the no-longer-so-secret secret menu has served to increase unpredictability. In its early years, these additional menu items were known only to regular customers, and the secret was shared among them by word of mouth. Because its secret menu is now public and the various items included in it are visible to all, In-N-Out Burger has turned its secret menu into something that is highly predictable.

In terms of calculability, the limited menu (even with the addition of the secret menu) makes it relatively easy for customers to estimate the cost of a meal. In addition, as is true of all fast-food chains, the prices are low and customers know (or at least believe) that they are going to get a lot of food for little money. The preparation and serving of food is fast, but not nearly as fast as at McDonald's since food is not prepared until it is ordered. For example, custom cooking of a burger with fresh ingredients can take as much as 12 minutes. Overall, there is less emphasis on quantitative factors at In-N-Out Burger and more on quality. This is reflected in the beef that was used in the company's early years:

> All hamburgers were made from fresh, 100 percent additive-, filler-, and preservative-free beef. . . . In-N-Out's butcher boned, hand-cut the chuck's front ribs and shoulder . . . no other part of the steer was ever used . . . , ground it up into beef, and molded it into hamburger patties before delivering them fresh to each store.[160]

Later, when the company and its facilities were much larger,

> specially selected cow and steer chucks arrived at the antiseptically clean commissary. The chain proudly proclaimed that it paid "a premium to purchase fresh, high-quality beef chucks" . . . To better enforce In-N-Out's quality standards, each chuck was inspected before being accepted. After In-N-Out's inspection, a team of skilled butchers boned and removed the meat.[161]

The careful way in which this is worded is interesting and revealing; we will return to it below.

There is control at In-N-Out, but it is not nearly as great as that at McDonald's. The drive-through gave, and continues to give, In-N-Out great control over customers as well as over workers because of their limited contact with customers. In at least one way, In-N-Out Burger has even more control than McDonald's. McDonald's has many franchises, and while systems are in place to exert control over them, franchisees do have a measure of autonomy and can act in ways that the corporation does not sanction. In contrast, In-N-Out Burger has refused to franchise its operations because it saw franchising "as a surefire path to losing quality control."[162] Since it owns all of its restaurants, In-N-Out has great control over them.

Thus, in various ways, In-N-Out Burger is McDonaldized, and because of that, as well as for other reasons, it has a number of irrationalities. For example, its great popularity and the enormous devotion of many of its customers have led to at least some inefficiencies such as long lines at the drive-through and in the restaurants. The use of fresh beef and raw potatoes can lead to more unpredictable results than the use of frozen beef and potatoes. It is far more difficult to control raw beef and fresh potatoes than frozen beef and potatoes. McDonald's gave up on fresh potatoes decades ago because of the mess and smell of rotting potato peels.

While In-N-Out Burger touts its quality, it is not clear that what it offers is truly high-quality products. Take, for example, the beef discussed above. We are told that the company pays a premium for the beef, and it is "high quality," "specially selected," inspected, and processed in antiseptically clean commissaries; it is hand cut; no additives, preservatives, or fillers are added to the hamburger; and the preformed hamburgers are delivered fresh to the restaurants. However, we are not told the key fact about the beef, that is, its USDA grade. Top-grade beef is prime or choice, but since the company does not use those terms to describe its beef, we can assume that it uses lower-grade beef. Furthermore, it admits to using beef from "cows" (females); the best beef comes from steers (young males), and no self-respecting steakhouse would ever sell cow meat.

While In-N-Out Burger in McDonaldized in various ways, what is most interesting about it is the degree to which it is not McDonaldized, or at least less McDonaldized than, say, McDonald's itself. Some of this is implied above. For example, the use of fresh beef and potatoes means that it must operate in a less rationalized manner than its competitors that use frozen beef and potatoes. For example, it is less efficient to ship and store fresh potatoes than those that are frozen. Hand-peeled and hand-cut potatoes that are fried raw will be less uniform than those cut by machine. Lettuce that is leafed by hand will not be as uniform as that which is handled industrially. In-N-Out Burger has relied more on humans and less on nonhuman technologies than most other chains.

In-N-Out Burger has worked hard to limit the McDonaldization of the work in its restaurants—to reduce the creation of McJobs. This is clear in the area of employee pay and benefits: "From the start, In-N-Out paid its employees more than the going rate (associates always made at least two to three dollars above minimum wage) and was an early practitioner of profit sharing . . . an expansive set of benefits under which part-time workers received free meals, paid vacations, 401(k) plans, and flexible schedules. Full-time associates also received medical, dental, vision, life and travel insurance."[163]

As a result of such policies, In-N-Out Burger had the lowest turnover rate in the fast-food industry. It is likely a bit of hyperbole, but Perman claims that In-N-Out Burger "was a place where people genuinely enjoyed getting up in the morning and going work."[164] No data are presented to support this view, and it is certainly the case that at least some employees didn't like going to work, didn't like the job, and quit because of that. Furthermore, it is not clear that the work itself is any more challenging or demanding than the work at McDonald's.

In-N-Out Burger's customers seem to like eating there more than those who eat at the other major chains. In fact, many customers seem quite devoted to it. Perman has a lot to say about this, but again she has no hard data; all she has is anecdotal evidence. Nonetheless, it does seem that In-N-Out Burger has more devoted customers than does McDonald's. Rather than simply having millions of customers, In-N-Out Burger has devoted fans, and it has arguably "been elevated to cult status."[165]

The reason for the loyalty and devotion of In-N-Out Burger's customers is traceable to a large extent to its ability to be, or at least appear to be, enchanted. There is a magic, a mystery, a "mystique"[166] about In-N-Out Burger that is not rivaled anywhere else in the fast-food industry. The best example of this is the chain's "secret menu," which, at first, was known only to "those in the know." Now, however, as was pointed out above, the secret has been made public (hence it is no longer a secret, if it ever really was). Nonetheless, the existence of this secret menu helped to enchant the chain, at least in its early years. By the way, the no longer secret menu is the following:

- Double Meat
- Grilled Cheese
- 3 × 3 (3 beef patties and 3 slices of cheese)
- 4 × 4 (same as 3 × 3, plus one more patty and cheese slice)
- Protein Style (burger wrapped in lettuce rather than a bun)
- Animal Style (a mustard cooked beef patty)

While there is some originality here (especially the last two menu items), half of the secret menu simply involves the supersizing found in many other fast-food chains.

Pret A Manger

Pret A Manger is British company, although the name is French, meaning "ready to eat." It was founded in London in 1968 and remains based primarily in Great Britain, although about 34 have opened in the United States (New York, Washington, D.C., and Chicago), and more are promised.[167] As of this writing, there are about 250 Pret A Manger restaurants in the world.[168] Unlike In-N-Out and McDonald's, Pret A Manger does *not* sell hamburgers. While it sells a variety of foods (salads, soups, wraps, desserts, etc.), Pret is best known for its high-quality sandwiches such as Balsamic Chicken & Avocado and Egg Salad, Parmesan, and Spinach. Like In-N-Out, Pret avoids the use of preservatives and chemicals. Pret does not make its sandwiches to order, but it does have them made in the shops several times a day. Those sandwiches and other products that have not sold at the end of the day are given away to charities.

While Pret strives in many ways to be the antithesis of McDonald's, a third of the company was sold to McDonald's about a decade ago. Following that sale, the company was pushed in the direction of the McDonald's model of great, perhaps over-, expansion, resulting in declining sales and closed restaurants. McDonald's sold its share in 2008, and the majority of the company is now owned by a private equity firm. Since then, Pret seems to have found its footing and niche once again and is expanding without losing the quality that makes it distinctive.

Nevertheless, Pret, like In-N-Out, has the basic characteristics of McDonaldization:

- There is an emphasis on efficiency and speed. The objective is to serve a customer within 60 seconds. While sandwiches are made fresh several times a day, they are wrapped and stored in cases where they can be grabbed by customers and eaten in the shop but more likely on the go. On the issue of speed, the company's chief executive said, "Pret a Manger does mean ready to eat—kapow!—not ready to wait."[169] The soups are not cooked by the "Hot Chef," but instead they "are sent to the shops premade and in plastic bags."[170]
- Control is exercised over the employees by mystery shoppers who visit the shops and evaluate the workers. In addition, a thick manual dictates what employees are supposed to do: "It states, for example, that employees should be 'bustling around and being active' on the floor, not 'standing around looking bored.' It encourages them to occasionally hand out free coffee or cakes to regulars, and not 'hide your true character' with customers."[171] Further control is exercised by the fact that after they have been working there for 3 months, employees must pass a quiz on basic Pret procedures. On the job, Hot Chefs do not set oven temperature controls by themselves, but rather the "ovens are preprogrammed for each baked item."[172] Control is exercised over customers in various ways. This is especially true in the

case of the sandwiches and the fact that they are premade, making it impossible for the customers in front of you in line to slow things down by asking for a little more or less of this and that.[173]

- Predictability is ensured by giving employees recipe cards detailing what should go into, and how to prepare, various foods. They also have access to pictures of how, for example, sandwiches should look when they are finished. Copying the picture leads to highly predictable sandwiches.

- Many aspects of Pret's operations are highly quantified. For example, in one case, a kitchen-supervisor-in-training had goals that included "six bowls of granola within 1 minute and 17 seconds; 24 edamame packs within 6 minutes and 2 seconds; 20 containers of honey-granola within 6 minutes and 17 seconds. Berry bowls, muesli bowls, porridge toppings—the list goes on. The food is taken to the shop floor as soon as it is finished."[174]

- Pret A Manger has its irrationalities, including the fact that much of its food is high in calories, salt, and sugar. While the emphasis on fresh and natural ingredients seems to imply that the food is also healthy, the fact is that many of the items sold at Pret are quite unhealthy. For example, a British newspaper revealed that the Posh Cheddar & Pickle Baguette "contains almost 800 calories and 15.6g of saturated fat . . . not dissimilar to a Big Mac and medium fries."[175]

While it is McDonaldized in various ways and to at least some degree, Pret, like In-N-Out, has various characteristics that seem to be at variance with the McDonald's model. Among other things, there is the high quality of much of its food, the high level of satisfaction among its employees and their comparatively low turnover rate, and the fact that Pret, like In-N-Out, has its devoted fans and has become something of a cult.

Overall, we are led to conclude that while there are important differences between In-N-Out and Pret and the highly McDonaldized chains like McDonald's, there are far more similarities than differences; they do adhere to the basic principles of McDonaldization. Thus, while In-N-Out and Pret are different in many ways, they are certainly *not* the antitheses of McDonald's and other highly McDonaldized systems.

A Look Ahead

Because this book is a work in the social sciences, it cannot merely assert that McDonaldization is spreading throughout society; it must present evidence for that assertion. Thus, after a discussion of the past, present, and future of McDonaldization in Chapter 2, Chapters 3 and 4 provide evidence of the four basic dimensions of McDonaldization outlined in this chapter: efficiency, calculability, predictability, and control. Numerous examples in each chapter show the degree to which McDonaldization has penetrated society and the accelerating rate of that penetration.

The remainder of the book is more analytical. In Chapter 5, the fifth and paradoxical element of McDonaldization—the irrationality of rationality—is explored. Although much of the book criticizes McDonaldization, this chapter presents the critique most clearly and directly, discussing a variety of irrationalities, the most important of which is dehumanization. In Chapter 6, individuals and groups bothered, if not enraged, by McDonaldization are offered ways of dealing with an increasingly McDonaldized world. Finally, Chapter 7 discusses the continued expansion of McDonaldization primarily under the heading of globalization. We will also discuss the possibility that a process of deMcDonaldization is under way in the form of seemingly contradictory processes of Starbuckization, eBayization, and the expansion of Web 2.0 (e.g., Facebook). We conclude that, while there is some evidence of deMcDonaldization, it is far from enough to allow us to conclude that McDonaldization is anywhere near its end. Rather, it leads us to a more nuanced sense of the McDonaldization process.

2

The Past, Present, and Future of McDonaldization

From the Iron Cage to the Fast-Food Factory and Beyond

McDonaldization did not emerge in a vacuum; it was preceded by a series of social and economic developments that not only anticipated it but also gave it many of the basic characteristics touched on in Chapter 1.[1] In the first half of this chapter, I will look briefly at a few of these developments. First, I will examine the notion of bureaucracy and Max Weber's theories about it and the larger process of rationalization. Next, I will offer a discussion of the Nazi Holocaust, a method of mass killing that can be viewed as the logical extreme of Weber's fears about rationalization and bureaucratization. Then, I will look at several intertwined socioeconomic developments that were precursors of McDonaldization: scientific management as it was invented at the turn of the century by F. W. Taylor, Henry Ford's assembly line, the mass-produced suburban houses of Levittown, the shopping mall, and Ray Kroc's creation of the McDonald's chain. These are not only of historical interest; most continue to be important to this day.

The second half of the chapter will shift to a focus on the present, as well as some thoughts on the future of McDonaldization. I begin with several of the forces driving McDonaldization today, including that it is profitable, we value it, and it fits with a range of other ongoing changes. Then I turn to the relationship between McDonaldization and three of the most important social changes of our time—the rise of postindustrial, post-Fordist, and postmodern society. The discussion then shifts to a much more specific issue, the ascent of Mt. Everest, because it seems to indicate that there are no limits to the future expansion of McDonaldization.

Bureaucratization: Making Life More Rational

A bureaucracy is a large-scale organization composed of a hierarchy of offices. In these offices, people have certain responsibilities and must act in accordance with rules, written regulations, and means of compulsion exercised by those who occupy higher-level positions.

The bureaucracy is largely a creation of the modern Western world. Although earlier societies had organizational structures, they were not nearly as effective as the bureaucracy. For example, in traditional societies, officials performed their tasks because of a personal loyalty to their leader. These officials were subject to personal whim rather than impersonal rules. Their offices lacked clearly defined spheres of competence, there was no clear hierarchy of positions, and officials did not have to obtain technical training to gain a position.

Ultimately, the bureaucracy differs from earlier methods of organizing work because of its formal structure, which, among other things, allows for greater efficiency. Institutionalized rules and regulations lead, even force, those employed in the bureaucracy to choose the best means to arrive at their ends. A given task is broken down into components, with each office responsible for a distinct portion of the larger task. Incumbents of each office handle their part of the task, usually following preset rules and regulations and often in a predetermined sequence. When each of the incumbents has, in order, handled the required part, the task is completed. By handling the task this way, the bureaucracy has used what its past history has shown to be the optimum means to the desired end.

Weber's Theory of Rationality

The roots of modern thinking on bureaucracy lie in the work of the turn-of-the-century German sociologist Max Weber.[2] His ideas on bureaucracy are embedded in his broader theory of the rationalization process. In the latter, Weber described how the modern Western world managed to become increasingly rational—that is, dominated by efficiency, predictability, calculability, and nonhuman technologies that control people. He also examined why the rest of the world largely failed to rationalize.

McDonaldization is an amplification and extension of Weber's theory of rationalization, especially into the realm of consumption. For Weber, the model of rationalization was the bureaucracy; for me, the fast-food restaurant is the paradigm of McDonaldization.[3]

Weber demonstrated in his research that the modern Western world had produced a distinctive kind of rationality. Various types of rationality had existed in all societies at one time or another, but none had produced the type that Weber called formal rationality. This is the sort of rationality I refer to when I discuss McDonaldization or the rationalization process in general.

According to Weber, formal rationality means that the search by people for the optimum means to a given end is shaped by rules, regulations, and larger social structures. Individuals are not left to their own devices in searching for the best means of attaining a given objective. Weber identified this type of rationality as a major development in the history of the world. Previously, people had been left to discover such mechanisms on their own or with vague and general guidance from larger value systems (religion, for example).[4] After the development of formal rationality, they could use institutionalized rules that help them decide—or even dictate to them—what to do. An important aspect of formal rationality, then, is that it allows individuals little choice of means to ends. In a formally rational system, virtually everyone can (or must) make the same, optimal choice.

Weber praised the bureaucracy, his paradigm of formal rationality, for its many advantages over other mechanisms that help people discover and implement optimum means to ends. The most important advantages are the four basic dimensions of rationalization (and of McDonaldization).

First, Weber viewed the bureaucracy as the most efficient structure for handling large numbers of tasks requiring a great deal of paperwork (now often computer work). As an example, Weber might have used the Internal Revenue Service (IRS), for no other structure could handle millions of tax returns as well.

Second, bureaucracies emphasize the quantification of as many things as possible. Reducing performance to a series of quantifiable tasks helps people gauge success. For example, an IRS agent is expected to process a certain number of tax returns each day. Handling less than the required number of cases is unsatisfactory performance; handling more is excellence. The quantitative approach presents a problem, however: little or no concern for the actual quality of work. Employees are expected to finish a task with little attention paid to how well it is handled. For instance, IRS agents who receive positive evaluations from their superiors for managing large numbers of cases may actually handle the cases poorly, costing the government thousands or even millions of dollars in uncollected revenue. Or the agents may handle cases so aggressively that taxpayers become angered.

Third, because of their well-entrenched rules and regulations, bureaucracies also operate in a highly predictable manner. Incumbents of a given office know with great assurance how the incumbents of other offices will behave. They know what they will be provided with and when they will receive it. Outsiders who receive the services that bureaucracies dispense know with a high degree of confidence what they will receive and when they will receive it. Again, to use an example Weber might have used, the millions of recipients of checks from the Social Security Administration know precisely when they will receive their money and exactly how much they will receive.

Finally, bureaucracies emphasize control over people through the replacement of human judgment with the dictates of rules, regulations, and structures. Employees are controlled by the division of labor, which allocates to each office a limited number of well-defined tasks. Incumbents must do those tasks, and no others, in the manner prescribed by the organization. They may not, in most cases, devise idiosyncratic ways of doing those tasks. Furthermore, by making few, if any, judgments, people begin to resemble human robots or computers. Bureaucracies can then consider replacing humans with machines. This replacement has already occurred to some extent: In many settings, computers have taken over bureaucratic tasks once performed by people. Similarly, the bureaucracy's clients are also controlled. They may receive appropriate services in certain ways and not others. For example, people can receive welfare payments by check or direct deposit, not in cash.

Irrationality and the "Iron Cage"

Despite the advantages it offers, bureaucracy suffers from the irrationality of rationality. Like a fast-food restaurant, a bureaucracy can be a dehumanizing place in which to work and by which to be served. Ronald Takaki characterizes rationalized settings as places in which the "self was placed in confinement, its emotions controlled, and its spirit subdued."[5] In other words, they are settings in which people cannot always behave as human beings—where people are dehumanized.

In addition to dehumanization, bureaucracies exhibit other irrationalities. Instead of remaining efficient, bureaucracies can become increasingly inefficient because of tangles of red tape and other pathologies. The emphasis on quantification often leads to large amounts of poor-quality work. Bureaucracies often become unpredictable as employees grow unclear about what they are supposed to do and clients do not get the services they expect. Because of these and other inadequacies, bureaucracies begin to lose control over those who work within and are served by them. Anger at the nonhuman technologies that replace them often leads employees to undercut or sabotage the operation of these technologies. All in all, what were designed as highly rational operations often end up being quite irrational.

Although Weber was concerned about the irrationalities of formally rationalized systems, he was even more animated by what he called the "iron (or steel) cage" of rationality. In Weber's view, bureaucracies are cages in the sense that people are trapped in them, their basic humanity denied. Weber feared most that bureaucracies would grow more and more rational and that rational principles would come to dominate an increasing number of sectors of society. He anticipated a society of people locked into a series of rational structures, who could move only from one rational system to another—from rationalized educational institutions to rationalized workplaces, from rationalized recreational settings to rationalized homes. Society would eventually become nothing more than a seamless web of rationalized structures; there would be no escape.

A good example of what Weber feared is found in the contemporary rationalization of recreational activities. Recreation can be thought of as a way to escape the rationalization of daily routines. However, over the years, these escape routes have themselves become rationalized, embodying the same principles as bureaucracies and fast-food restaurants. Among the many examples of the rationalization of recreation[6] are cruises and cruise lines,[7] chains of campgrounds, and package tours. Take, for example, a 7-day Mediterranean cruise. The ship sails around at least a part of the Mediterranean, stopping briefly at major tourist attractions and towns along the coast of, say, southern Europe. This route allows tourists to glimpse the maximum number of sites in the 7-day period. At particularly interesting or important sights, the ship docks for a few hours to allow individuals to debark, have a quick local meal, buy souvenirs, and take some pictures. Then a quick trip back to the ship, and it is off to the next locale. The cruise goers sleep during the overnight trips to these locales and take most of their meals on board ship. They awaken the next morning, have a good breakfast, and there they are at the next site. It's all very efficient. With the rationalization of even their recreational activities, people do come close to living in Weber's iron cage of rationality.

The Holocaust: Mass-Produced Death

Weber wrote about the iron cage of rationalization and bureaucratization in the early 1900s. Zygmunt Bauman argues that Weber's worst fears about these processes were realized in the Nazi Holocaust, which began within a few decades of Weber's death in 1920.

Bauman contends that "the Holocaust may serve as a paradigm of modern bureaucratic rationality."[8] Like bureaucracy, the Holocaust was a distinctive product of Western civilization. In fact, Bauman argues that the Holocaust was not an aberration but "in keeping with everything we know about our civilization, its guiding spirit, its priorities, its immanent vision of the world."[9] That is, the Holocaust required the rationality of the modern world. It could not have occurred in premodern, less rationalized societies.[10] In fact, the pogroms that had occurred in premodern societies were too inefficient to allow for the systematic murder of millions of people that occurred in the Holocaust.

The Holocaust can also be seen as an example of modern social engineering in which the goal was a perfectly rational society. To the Nazis, a perfect society was one free of Jews, as well as of gypsies, gays, lesbians, and the disabled. Hitler himself defined the Jews as a "virus," a disease that had to be eliminated from Nazi society.

The Holocaust had all the basic characteristics of rationalization (and McDonaldization). It was an effective mechanism for the destruction of massive numbers of human beings. For example, early experiments showed that bullets were inefficient; the Nazis eventually settled on gas as the most efficient means of destroying people. The Nazis also found it efficient to use members of the Jewish community to perform a variety of tasks (for example, choosing the next group of victims) that the Nazis otherwise would have had to perform themselves.[11] Many Jews cooperated because it seemed like the "rational" thing to do (they might be able to save others or themselves) in such a rationalized system.

The Holocaust emphasized quantity, such as how many people could be killed in the shortest time.[12] There was certainly little attention paid to the quality of the life, or even of the death, of the Jews as they marched inexorably to the gas chambers.

In another quantitative sense, the Holocaust has the dubious distinction of being seen as the most extreme of mass exterminations:

> Like everything else done in the modern—rational, planned, scientifically informed, expert, efficiently managed, coordinated—way, the Holocaust . . . put to shame all its alleged pre-modern equivalents, exposing them as primitive, wasteful and ineffective by comparison . . . the Holocaust . . . towers high above the past genocidal episodes.[13]

The Holocaust involved an effort to make mass murder routine. The whole process had an assembly-line quality about it. Trains snaked their way toward the concentration camps; victims lined up and followed a set series of steps. Once the process was complete, camp workers produced stacks of dead bodies for systematic disposal.

Finally, the victims of the Holocaust were managed by a huge nonhuman technology:

> [Auschwitz] was also a mundane extension of the modern factory system. Rather than producing goods, the raw material was human beings and the end-product was death, so many units per day marked carefully on the manager's production charts. The chimneys, the very symbol of the modern factory system, poured forth acrid smoke produced by burning human flesh. The brilliantly organized railroad grid of modern Europe carried a new kind of raw material to the factories. It did so in the same manner as with other cargo. . . . Engineers designed the crematoria; managers designed the system of bureaucracy that worked with a zest and efficiency.[14]

Needless to say, the Holocaust represented the ultimate in the irrationality of rationality. After all, what could be more dehumanizing than murdering millions of people in such a mechanical way? Furthermore, for the murders to have occurred in the first place, the victims had to be dehumanized—that is, "reduced to a set of quantitative measures."[15] Overall, "German bureaucratic machinery was put in the service of a goal incomprehensible in its irrationality."[16]

Discussing the Holocaust in the context of McDonaldization may seem extreme to some readers, perhaps even unreasonable. Clearly, the fast-food restaurant cannot be discussed in the same breath as the Holocaust. There has been no more heinous crime in the history of humankind. Yet I have strong reasons for presenting the Holocaust as a precursor of McDonaldization. First, the Holocaust was organized around the principles of formal rationality, relying extensively on the paradigm of that type of rationality—the bureaucracy. Second, the Holocaust was also linked to the factory system, which you will soon discover was related to other precursors of McDonaldization. Finally, the spread of formal rationality today, through the process of McDonaldization, supports Bauman's view that something like the Holocaust could happen again.

Scientific Management: Finding the One Best Way

A less dramatic but no less important precursor to McDonaldization was the development of scientific management. In fact, Weber at times mentioned scientific management in his discussion of the rationalization process.

Scientific management was created by Frederick W. Taylor in the late 19th and early 20th centuries. His ideas played a key role in shaping the work world throughout the 20th century.[17] Taylor developed a series of principles designed to rationalize work and was hired by a number of large organizations (for example, Bethlehem Steel) to implement those ideas, mostly in their factories.

Taylor was animated by the belief that the United States suffered from "inefficiency in almost all our daily acts" and that there was a need for "greater national efficiency"; his followers came to be known as "efficiency experts." His "time-and-motion" studies were designed to replace what Taylor called the inefficient "rule-of-thumb" methods, which dominated work in his day, with what he thought of as the "one best way"—that is, the optimum means to the end of doing a job.[18] Taylor outlined a series of steps to be followed in time-and-motion studies, including finding skilled workers; studying their elementary movements (as well as their tools and implements); timing each step carefully (calculability) with the aim of discovering the most efficient way of accomplishing each of them; eliminating "all false movements, slow movements, and useless movements"; and finally combining the most efficient movements (and tools) to create the "one best way" of doing a job.[19]

Scientific management also placed great emphasis on predictability. Clearly, in delineating the one best way to do a job, Taylor sought an approach that each and every worker could use. Taylor also believed that allowing workers to choose their own tools and methods of doing a job led to low productivity and poor quality. Instead, he sought the complete standardization of tools and work processes. Taylor also favored clear and detailed standards that made sure all workers did a given type of job in the same way and would therefore consistently produce high-quality work.

Overall, scientific management produced a nonhuman technology that exerted great control over workers. When workers followed Taylor's methods, employers found that they worked much more efficiently, that everyone performed the same steps (that is, their work exhibited predictability), and that they produced a great deal more while their pay had to be increased only slightly (another instance of emphasizing calculability). Taylor's methods thus meant increased profits for those enterprises that adopted them.

Like all rational systems, scientific management had its irrationalities. Above all, it was a dehumanizing system in which people were considered expendable and treated as such. Furthermore, because workers did only one or a few tasks, most of their skills and abilities remained unused.

Although one hears little these days of Taylor, efficiency experts, and time-and-motion studies, their impact is strongly felt in a McDonaldized society. For instance, hamburger chains strive to discover and implement the "one best way" to grill hamburgers, cook French fries, prepare shakes, process customers, and the rest. The most efficient ways of handling a variety of tasks have been codified in training manuals and taught to managers who, in turn, teach them to new employees. The design of the fast-food restaurant and its various technologies have been put in place to aid in the attainment of the most efficient means to the end of feeding large numbers of people. Here, again, McDonald's did not invent these ideas but, rather, brought them together with the principles of the bureaucracy and of the assembly line, thus contributing to the creation of McDonaldization.

The Assembly Line: Turning Workers Into Robots

Like modern bureaucracy and scientific management, the assembly line came into existence at the dawn of the 20th century. Pioneered in the bureaucratized automobile industry, the ideas of scientific management helped shape that industry. The automobile assembly line was invented mainly because Henry Ford wanted to save time, energy, and money (that is, to be more efficient). Greater efficiency would lead to lower prices, increased sales, and greater profitability for the Ford Motor Company.[20]

Ford got the idea for the automobile assembly line from the overhead trolley system used at the time by Chicago meatpackers to butcher cattle. As the steer was propelled along on the trolley system, a line of highly specialized butchers performed specific tasks so that, by the end of the line, the steer had been completely butchered. This system was clearly more efficient than having a single meat cutter handle all these tasks.

On the basis of this experience and his knowledge of the automobile business, Ford developed a set of principles for the construction of an automobile assembly line, principles that to this day stand as models of efficiency, including reducing work-related movements to an absolute minimum, parts needed in the assembly process are to travel the least possible distance, mechanical (rather than human) means are to be used to move the car (and parts) from one step in the assembly process to the next, and complex sets of movements are to be eliminated with the worker doing "as nearly as possible only one thing with one movement."[21]

The Japanese adopted American assembly-line technology after World War II and then made their own distinctive contributions to heightened efficiency. For example, the Japanese "just-in-time" system replaced the American "just-in-case" system. Both systems refer to the supply of needed parts to a manufacturing operation. In the American system, parts are stored in the plant until, or in case, they are needed. This system leads to inefficiencies such as the purchase and storage (at great cost) of parts that will not be needed for quite some time. To counter these inefficiencies, the Japanese developed the just-in-time system: Needed parts arrive at the assembly line just as they are to be placed in the car (or whatever object is being manufactured). In effect, all the Japanese company's suppliers become part of the assembly-line process.

In either system, the assembly line permits the quantification of many elements of the production process and maximizes the number of cars or other goods produced. What each worker on the line does, such as putting a hubcap on each passing car, is highly predictable and leads to identical end products.

The assembly line is also a nonhuman technology that permits maximum control over workers. It is immediately obvious when a worker fails to perform the required tasks. There would, for example, be a missing hubcap as the car moves down the line. The limited time allotted for each job allows little or no room for innovative ways of doing a specific task. Thus, fewer, as well as less-skilled, people are able to produce cars. Furthermore, the specialization of each task permits the replacement of human workers with robots. Today, mechanical robots handle more and more assembly-line tasks.

As has been well detailed by many observers, the assembly line carries with it much irrationality. For example, it can be a dehumanizing setting in which to work. Human beings, equipped with a wide array of skills and abilities, are asked to perform a limited number of highly simplified tasks over and over. Instead of expressing their human abilities on the job, people are forced to deny their humanity and to act like robots.

Despite its flaws, the assembly line represented a remarkable step forward in the rationalization of production and became widely used throughout manufacturing. Like bureaucracy and even the Holocaust, the automobile assembly line is an excellent illustration of the basic elements of formal rationality.

The assembly line also has had a profound influence on the development of the fast-food restaurant. "The people who pioneered fast food revered Ford's assembly-line methods."[22] The most obvious example of this mimicry is the conveyor belt used by Burger King to cook its hamburgers. Less obvious is the fact that much of the work in a fast-food restaurant is performed in assembly-line fashion, with tasks broken down into their simplest components. For example, "making a hamburger" means grilling the burgers, putting them on the rolls, smearing on the "special sauce," laying on the lettuce and tomato, and wrapping the fully dressed burgers. Even customers face a kind of assembly line, the drive-through window being the most obvious example. As one observer notes, "The basic elements of the factory have obviously been introduced to the fast-food phenomenon . . . [with] the advent of the feeding machine."[23]

In addition to being a precursor, the automobile assembly line laid the groundwork for McDonaldization in another way. Mass production gave many people ready access to affordable automobiles, which in turn led to the immense expansion of the highway system and the tourist industry that grew up alongside it.[24] Restaurants, hotels, campgrounds, gas stations, and the like arose and served as the precursors to many of the franchises that today lie at the heart of the McDonaldized society.[25]

Levittown: Putting Up Houses—"Boom, Boom, Boom"

The availability of the automobile helped make possible not only the fast-food restaurant but also suburbia, especially the mass-produced suburban houses pioneered by Levitt & Sons. Between 1947 and 1951, this company built 17,447 homes on former New York potato fields, thereby creating Levittown, Long Island, and an instant community of 75,000 people.[26] The first houses in the planned community of Levittown, Pennsylvania, went on sale in 1958. The two Levittowns provided the model for innumerable contemporary suburban developments. With their need for and access to automobiles, suburban dwellers were, and are, a natural constituency for the fast-food restaurant.

Levitt & Sons thought of their building sites as large factories using assembly-line technology: "What it amounted to was a reversal of the Detroit assembly line. . . . There, the car moved while the workers stayed at their stations. In the case of our houses, it was the workers who moved, doing the same jobs at different locations."[27]

The workers performed specialized tasks, much like their compatriots on the automobile assembly line: "The same man does the same thing every day, despite the psychologists. It is boring; it is bad; but the reward of the green stuff seems to alleviate the boredom of the work."[28] The Levitts thus rationalized the work of the construction laborer much as Ford had done with the automobile worker, with much the same attitude toward the worker.

The housing site as well as the work was rationalized. In and around the building locale, the Levitts constructed warehouses, woodworking shops, plumbing shops, and a sand, gravel, and cement plant. Thus, instead of buying these services and their resulting products from others and then shipping them to the construction site, the products and services were onsite and controlled by the Levitts. Where possible, the Levitts also used prefabricated products. They deemed manufacturing an entirely prefabricated house less efficient, however, than making a partially prefabricated one.

The actual construction of each house followed a series of rigidly defined and rationalized steps. For example, in constructing the wall framework, the workers did no measuring or cutting; each piece had been cut to fit. The siding for a wall consisted of 73 large sheets of Colorbestos, replacing the former requirement of 570 small shingles. All houses were painted under high pressure, using the same two-tone scheme—green on ivory. As a result, "Once the groundwork is down, houses go up boom, boom, boom."[29] The result, of course, was a large number of nearly identical houses produced quickly at low cost.

The emphasis on quantitative factors went beyond the physical construction of the house. Advertisements for Levittown houses stressed "the size and value of the house."[30] In other words, Levittown, like its many successors in the march toward increased rationalization, tried to convince consumers that they were getting the most for the least money.

These principles, once used exclusively in low-priced homes, have since been applied to high-priced homes, as well. "McMansions" are increasingly often little more than huge and luxuriously appointed, factory-made modular homes.[31]

Many have criticized life in identical houses in highly rationalized communities. One early critique renamed suburbia "Disturbia," describing the suburban home as a "split-level trap."[32] However, one can also look positively at suburban rationalization. For example, many residents of Levittown have customized their homes so that they no longer look as homogeneous as before.[33] Other observers have found much of merit in Levittown and suburbia. Herbert Gans, for example, concluded his study of a third Levittown built in New Jersey by arguing that "whatever its imperfections, Levittown is a good place to live."[34] Whether or not it is a "good" place to live, Levittown is certainly a rationalized place.

Shopping Centers: Malling America

Another component of rationalized society whose development was fueled by the rise of automobiles, and of suburban housing, was the fully enclosed shopping mall.[35] The modern mall had precursors in arcades such as the Galleria Vittorio Emanuele in Milan, Italy (completed in 1877), and the first planned outdoor shopping center in the United States (built in 1916 in Lake Forest, Illinois). The original fully enclosed shopping mall, however, was Southdale Center in Edina, Minnesota, which opened in 1956, not long after the opening of Ray Kroc's first McDonald's. Today, tens of thousands of malls in the United States are visited by hundreds of millions of shoppers each month. The largest shopping mall in the United States to date, the Mall of America, opened in 1992 down the road from Edina, in Bloomington, Minnesota. It includes four department stores, 520 specialty shops (many of them parts of chains), and an amusement park.[36] Enormous malls have become a global phenomenon. In fact, 8 of the 10 largest mega-malls are now found in Asia.[37] The largest, by far, is the South China Mall in Dongguan, China, which dwarfs the Canadian and American mega-malls. It has 7.1 million square feet compared to 3.8 million at the West Edmonton Mall and 2.8 million at the Mall of America. The South China Mall has 1,500 stores, West Edmonton 800, and Mall of America 520. While most of the world's largest mega-malls are now in China, others can be found in the Philippines, Dubai, Turkey, Kuala Lumpur, Colombia, Thailand, and Brazil. Found in these mega-malls are themed zones, ice skating rinks, a walk-through aquarium, a 48-story hotel, and so on.[38]

Shopping malls and McDonaldized chains complement each other beautifully. The malls provide a predictable, uniform, and profitable venue for such chains. For their part, most malls would have much unrented space and would not be able to exist were it not for the chains. Simultaneous products of the fast-moving automobile age, malls and chains feed off each other, furthering McDonaldization.

Ironically, malls today have become a kind of community center for both young and old. Many elderly people now use malls as places to both exercise and socialize. Teens prowl the malls after school and on weekends, seeking social contact and checking out the latest in fashions and mass entertainment. Because some parents also take their children to malls to "play," malls are now offering play rooms (free ones as well as profit-making outlets that charge an entry fee and offer things like free video games and free movies).[39] Like many other contributors to the McDonaldization of society, malls strive to engage customers from the cradle to the grave.

William Kowinski argues that the mall "was the culmination of all the American dreams, both decent and demented; the fulfillment, the model of the postwar paradise."[40] One could give priority to the mall, as Kowinski does, and discuss the "malling of America." However, in my view, the fast-food restaurant is a far more powerful and influential force. Like the mall, however, McDonaldization can be seen as both "decent and demented."

It is important to note that since the onset of the "Great Recession" in late 2007, shopping malls throughout the United States have been plagued by an increasing number of empty shops (a devastating sight in malls that adversely affects the spectacle they are endeavoring to create), and many of these shops are outlets of chains that are retrenching or themselves going out of business.[41] Closed shops mean less revenue to the mall owners and make it more difficult for them to survive. Many malls are themselves in danger of closing, are in the process of closing, or have already closed as a result of the recession.[42] The result is the proliferation of "dead malls" (see www.deadmalls.com) such as the Randall Park Mall outside Cleveland, Ohio.[43] However, strip malls and even fully enclosed "big box" malls have been experiencing problems for some time.

McDonald's: Creating the "Fast-Food Factory"

The basic McDonald's approach was created by two brothers, Mac and Dick McDonald, who opened their first restaurant in Pasadena, California, in 1937.[44] They based the restaurant on the principles of high speed, large volume, and low price. To avoid chaos, they offered customers a highly circumscribed menu. Instead of personalized service and traditional cooking techniques, the McDonald brothers used assembly-line procedures for cooking and serving food. In place of trained cooks, the brothers' "limited menu allowed them to break down food preparation into simple, repetitive tasks that could be learned quickly even by those stepping into a commercial kitchen for the first time."[45] They pioneered the use of specialized restaurant workers such as "grill men," "shake men," "fry men," and "dressers" (those who put the "extras" on burgers and who wrap them). They developed regulations dictating what workers should do and even what they should say. In these and other ways, the McDonald brothers took the lead in developing the rationalized "fast-food factory."[46]

Ray Kroc invented neither the McDonald's principles nor the idea of a franchise. "Franchising is a system in which one large firm . . . grants or sells the right to distribute its products or use its trade name and processes to a number of smaller firms. . . . Franchise holders, although legally independent, must conform to detailed standards of operation designed and enforced by the parent company."[47] The Singer Sewing Machine company pioneered franchising after the Civil War, and automobile manufacturers and soft drink companies were using it by the turn of the 20th century. By the 1930s, it had found its way into retail businesses such as Western Auto, Rexall Pharmacy, and the IGA food markets.

In fact, there had been many efforts to franchise food service before Kroc arrived on the scene in the early 1950s. The first food service franchises, the A&W Root Beer stands, made their debut in 1924. Howard Johnson began franchising ice cream and other food in 1935. The first Dairy Queen opened in 1944; efforts to franchise it nationally led to a chain of about 2,500 outlets by 1948. Other well-known food franchises predated McDonald's. Big Boy started in the late 1930s, and Burger King (then InstaBurger) and Kentucky Fried Chicken began in 1954. Thus, Kroc's first McDonald's, which opened on April 15, 1955, was a relative latecomer to the franchising business in general and the food franchise business in particular.

In 1954, when Ray Kroc first visited it, McDonald's was a single drive-in hamburger stand in San Bernardino, California (ironically, the same city where Taco Bell was founded by Glen Bell[48]). The basic menu, the approach, and even some of the techniques that McDonald's is famous for today had already been created by the McDonald brothers. Although it was a local sensation, the McDonald brothers were content to keep it that way; they were doing very well and had few grand ambitions despite a few tentative steps toward franchising. With plenty of ambition for all of them, Kroc became their franchising agent and went on to build the McDonald's empire of franchises. At first, Kroc worked in partnership with the McDonald brothers, but after he bought them out in 1961 for $2.7 million, he was free to build the business as he wished.

Kroc took the specific products and techniques of the McDonald brothers and combined them with the principles of other franchises (food service and others), bureaucracies, scientific management, and the assembly line. Kroc's genius was in bringing all these well-known ideas and techniques to bear on the fast-food business and adding his ambition to turn it, through franchising, into a national, then international, business. McDonald's and McDonaldization, then, do not represent something new but, rather, the culmination of a series of rationalization processes that had been occurring throughout the 20th century.

Kroc's major innovation lay in the way he franchised McDonald's. He did not permit regional franchises in which a single franchisee received control over all the outlets to be opened in a given area. Other franchisers had foundered because regional franchisees had grown too powerful and subverted the basic principles of the company. Kroc maximized central control, and thereby uniformity throughout the system, by granting franchises one at a time and rarely granting more than one franchise to a specific individual. Kroc also gained control over, and profited from, the franchisee's property.[49] Another of Kroc's innovations was to set the fee for a franchise at a rock-bottom $950. Other franchisers had set very high initial fees and made most of their money from the initial setup. As a result, they tended to lose interest in the continued viability of the franchisees. At McDonald's, profits did not come from high initial fees but from the 1.9% of store sales that headquarters demanded of its franchisees. The success of Kroc and his organization thus depended on the prosperity of the franchisees. This mutual interest was Kroc's greatest contribution to the franchise business and a key factor in the success of McDonald's and its franchisees, many of whom became millionaires.

Although Kroc imposed and enforced a uniform system, he encouraged the franchisees to come up with innovations that could enhance not only their operations but also those of the system as a whole. Take the case of product innovations. Kroc himself was not a great product innovator. One of his most notorious flops was the Hulaburger, a slice of grilled pineapple between two pieces of cheese wrapped in a toasted bun. Successful creations, such as the fish sandwich (the Filet-o-Fish), the Egg McMuffin, McDonald's breakfast meals, and even the Big Mac, came from franchisees. Thus, McDonald's achieved a balance between centralized control and the independence of franchisees.

Kroc spearheaded a series of developments that further rationalized the fast-food business.[50] For one thing, he (unwittingly) served as preacher and cheerleader for the principles of rationalization as he lectured "about uniformity, about a standardized menu, one size portions, same prices, same quality in every store."[51] This uniformity allowed McDonald's to differentiate itself from its competitors, whose food was typically inconsistent. McDonald's also led the field by imposing a limited menu (at first, 10 items), by creating tough standards for the fat content of its hamburgers, by converting to frozen hamburgers and French fries, by using inspectors to check on uniformity and conformity, and by forming in 1961 the first full-time training center in the business (called Hamburger University and offering a "degree" in "hamburgerology"). Today, more than 80,000 managers in McDonald's restaurants have graduated from Hamburger University (and 5,000 students attend each year), now located in a

130,000-square-foot, state-of-the-art facility on the McDonald's Home Office Campus in Oak Brook, Illinois. Because of McDonald's international scope, translators and electronic equipment enable professors to teach and communicate in 28 languages at one time. McDonald's also manages 10 international training centers, including Hamburger Universities in England, Japan, Germany, Brazil, Hong Kong, and Australia.[52]

In 1958, McDonald's published an operations manual that detailed how to run a franchise.[53] This manual laid down many of the principles for operating a fast-food restaurant:

> It told operators *exactly* how to draw milk shakes, grill hamburgers, and fry potatoes. It specified *precise* cooking times for all products and temperature settings for all equipment. It fixed *standard* portions on every food item, down to the *quarter ounce* of onions placed on each hamburger patty and the *thirty-two slices per pound* of cheese. It specified that French fries be cut at *nine thirty-seconds of an inch* thick. And it defined quality *controls* that were unique to food service, including the disposal of meat and potato products that were held more than *ten minutes* in a serving bin.
>
> . . . Grill men . . . *were instructed* to put hamburgers down on the grill moving from left to right, creating *six rows of six* patties each. And because the first two rows were farthest from the heating element, they were instructed (and still are) to flip the third row first, then the fourth, fifth, and sixth before flipping the first two. (italics added)[54]

It is hard to imagine a more rational system. Given this history, we now turn to the contemporary status, and some future possibilities, of McDonald's and McDonaldization.

McDonaldization and Contemporary Social Changes

The Forces Driving McDonaldization: It Pays, We Value It, It Fits

The attractiveness of the principles of McDonaldization, as well as its many advantages, help to account for its contemporary success and spread, but three other factors are also important in understanding its increasing prevalence: (1) material interests, especially economic goals and aspirations; (2) the culture of the United States, which values McDonaldization as an end in itself; and (3) the degree to which McDonaldization is attuned to important changes taking place within society.

Higher Profits and Lower Costs

Profit-making enterprises pursue McDonaldization because it leads to lower costs and higher profits. Clearly, greater efficiency and increased use of nonhuman technology are often implemented to increase profitability. Greater predictability provides, at the minimum, the climate needed for an organization to be profitable and for its profits to increase steadily from year to year. An emphasis on calculability, on things that can be quantified, helps lead to decisions that can produce and increase profits and makes possible measurements of profitability. In short, people and organizations profit greatly from McDonaldization, and as a result, they aggressively seek to extend its reach.

Although not oriented to profits, nonprofit organizations also press McDonaldization for material reasons. Specifically, it leads to lower costs, which permit nonprofit agencies, often in tight economic circumstances, to remain in operation and perhaps even to expand.

McDonaldization for Its Own Sake

Although economic factors lie at the root of McDonaldization, it has become such a desirable process that many people and enterprises pursue it as an end in itself. Many have come to value efficiency, calculability, predictability, and control and seek them out whether or not economic gains will result. For example, eating in a fast-food restaurant or having a microwave dinner at home may be efficient, but it is more costly than preparing the meal "from scratch."[55] Because they value efficiency, people are willing to pay the extra cost.

Americans have long valued rationalization, efficiency, and so on, and McDonald's has simply built on that value system. Furthermore, since its proliferation in the late 1950s, McDonald's (to say nothing of the myriad other agents of McDonaldization) has invested enormous amounts of money and great effort in convincing people of its value and importance. Indeed, it now proclaims itself a part of a rich American tradition rather than, as many people believe, a threat to it. Many Americans have eaten at McDonald's in their younger years, gone out with teenage buddies for a burger, taken their children there at various times as they grew up, or gone there to have a cup of coffee with their parents. McDonald's has exploited such emotional baggage to create a large number of highly devoted customers. Even though McDonald's is built on rational principles, its customers' loyalty is, at least to some degree, emotional. Thus, McDonaldization is likely to proceed apace for two reasons: It offers the advantages of rationality, and people are committed to it emotionally. This commitment is what leads people to ignore the company's disadvantages; this acceptance, in turn, helps open the world to even further advances in McDonaldization.

While adults may have some emotional commitment to McDonald's, this attachment is even more true of children.[56] In fact, children care little about the rational aspects of McDonald's; they are drawn to it because of emotions created to a large extent by advertisements. Children are more likely to value McDonald's for its own sake and not because it seems to offer material, especially economic, advantages.

Even new entrepreneurs may be attracted to opening a franchise because it has become such an attractive and popular method of operating a business. In so doing, they may ignore the fact that a given franchise market is saturated and there is little chance of making a profit in such a market.

McDonaldization and the Changing Society

A third explanation of the rush toward McDonaldization is that it meshes well with other changes occurring in American society and around the world. For example, the number of single-parent families and the number of women working outside the home have increased greatly. There is less likely to be anyone with the time to shop, prepare the ingredients, cook the food, and clean up afterward. There may not even be time (or money), at least during the work week, for meals at traditional restaurants. The speed and efficiency of a fast-food meal fits in well with these realities. Many other McDonaldized institutions offer similar matches.

The fast-food model also thrives in a society that emphasizes mobility, especially by automobile. Teenagers and young adults in the United States (and elsewhere), the most likely devotees of the fast-food restaurant, now have ready access to automobiles. And they need automobiles to frequent most fast-food restaurants, except those found in the hearts of large cities.

More generally, the fast-food restaurant suits a society in which people prefer to be on the move.[57] Going out for a McDonaldized dinner (or any other rationalized activity) is in tune with the demands of such a society; even better is the use of the drive-through window so that people do not even have to stop to eat. Further serving McDonaldization is the increasing number of people who travel about, either on business or for vacations. People on the move seem to like the idea that, even though they are in a different part of the country (or the world), they can still go to a familiar fast-food restaurant to eat the same foods they enjoy at home.

The increasing affluence of at least a portion of the population, accompanied by more discretionary funds, is another factor in the success of fast-food restaurants. People who have extra funds can support a fast-food "habit" and eat at such restaurants regularly. At the same time, the fast-food restaurant offers poor people the possibility of an occasional meal out.

The increasing influence of the mass media also contributes to the success of fast-food restaurants. Without saturation advertising and the ubiquitous influence of television and other mass media, fast-food restaurants would not have succeeded as well as they have. Similarly, the extensive advertising employed by McDonaldized systems has helped make them resounding successes.

Of course, technological change has probably played the greatest role in the success of McDonaldized systems. Initially, technologies such as bureaucracies, scientific management, the assembly line, and the major product of that production system, the automobile, all contributed to the birth of the fast-food society. Over the years, innumerable technological developments have both spurred, and been spurred by, McDonaldization: automatic drink dispensers, supermarket scanners, foods that cook themselves, the microwave oven, the 24-second clock in professional basketball, ATMs, voice mail, OnStar navigation systems, laptops, iPods, PDAs, and many others. Many technological marvels of the future will either arise from the expanding needs of a McDonaldizing society or help create new areas to be McDonaldized.

Today, the computer is the technology that contributes the most to the growth of McDonaldization.[58] In particular is the burgeoning importance of the Internet. Internet technology such as portals (for example, Internet Explorer) and search engines (such as Google and Yahoo!) have greatly rationalized and simplified access to the Internet. Today, the Internet is user-friendly and accessible to billions of people who are largely ignorant of computer technology and computer programming.[59]

All of the above factors are associated with aspects of the contemporary world associated with "modernity" and the "modern" world. There are many who argue, however, that we have moved, or are in various ways moving, beyond a modern world to one that is described as postindustrial, post-Fordist, and postmodern.

Other Major Social Changes: McDonaldization in the Era of the "Posts"

The ideas to be discussed in this section imply that "modern" phenomena such as McDonaldization are likely to soon disappear. I maintain, however, that McDonaldization and its modern (as well as its industrial and Fordist) characteristics are not only here for the foreseeable future but are also influencing society at an accelerating rate. Although important postindustrial, post-Fordist, and postmodern trends are also occurring, some thinkers associated with these perspectives have been too quick to declare an end to modernity, at least in its McDonaldized form.

Postindustrialism and McDonaldization:
"Complexification" and "Simplification"

Daniel Bell (and many others) argues that we have moved beyond industrial society to a new, postindustrial society.[60] Among other things, this change means that the focus in society has shifted from producing goods to providing services. Throughout most of the 20th century, the production of goods such as steel and automobiles dominated the economy of the United States. Today, however, the economy is dominated by services such as those related to education, computers, health care, and fast food. The rise of new technologies and the growth in knowledge and information processing are also characteristic of postindustrial society. Professionals, scientists, and technicians have increased in number and importance. The implication is that postindustrial society will be dominated by creative knowledge workers and not by routinized employees of McDonaldized systems.

However, the low-status service occupations that are so central to a McDonaldized society show no sign of disappearing. In fact, they have expanded. McDonaldization is built on many of the ideas and systems of industrial society, especially bureaucratization, the assembly line, and scientific management. Society is certainly postindustrial in many ways, and knowledge workers have grown more important, but the spread of McDonaldization indicates that some aspects of industrial society are still with us and will remain so for some time to come.

Jerald Hage and Charles Powers contend that a new postindustrial organization has arisen and coexists with the classic industrial organization, as well as with other organizational forms.[61] The postindustrial organization has a number of characteristics, including a leveling of hierarchical distinctions, a blurring of boundaries between organizations, a more integrated and less specialized organizational structure, an increase in behavior that is not bound by rules, and hiring policies that emphasize the creativity of potential employees. In contrast, McDonaldized organizations continue to be hierarchical, the behavior of employees and even managers is tightly bound by rules, and the last thing on the minds of those hiring for most jobs is creativity. Hage and Powers contend that jobs involving "tasks that are most clearly defined, technically simple, and most often repeated" are being eliminated by automation.[62] While many such jobs have been eliminated in heavy industry, they are not only alive and well but growing in McDonaldized service organizations.[63] Postindustrial organizations are also characterized by customized work and products, whereas standardized work (everyone follows the same procedures, scripts) and uniform products are the norm in McDonaldized settings. Unquestionably, postindustrial organizations are on the ascent, but McDonaldized organizations are also spreading. Modern society is sustaining contradictory organizational developments.

Hage and Powers envision a broader change in society as a whole. The emphasis will be on creative minds, complex selves, and communication among people who have these characteristics. They argue that "complexification will be the prevailing pattern of social change in postindustrial society."[64]Although some aspects of modern society are congruent with that image, McDonaldization demands uncreative minds, simple selves, and minimal communication dominated by scripts and routines. McDonaldization emphasizes "simplification," not "complexification."

In sum, the postindustrial thesis is not wrong but is more limited than many of its adherents believe. Postindustrialization coexists with McDonaldization. My view is that both complexification and simplification will prevail but in different sectors of the economy and the larger society.

Fordism and Post-Fordism: Or Is It McDonaldism?

A similar issue concerns a number of thinkers, especially those associated with Marxism, who claim that industry has undergone a transition from Fordism to post-Fordism. Fordism, of course, refers to the ideas, principles, and systems spawned by Henry Ford and discussed previously.

Fordism has a number of characteristics, including *mass production of a homogeneous product, inflexible technologies, standardized work routines* (or Taylorism), *efforts to increase productivity,* and a *market for mass-produced items.* Although Fordism grew throughout the 20th century, especially in the heavy industry of the United States, it reached its peak and began to decline in the 1970s. The oil crisis of 1973 and the subsequent downturn in the American automobile industry were primary factors in the decline of Fordism.

Some argue that the decline of Fordism has been accompanied by the rise of post-Fordism, which has a number of its own distinguishing characteristics:

- *Declining interest in mass products and growing interest in more customized and specialized products.* Style and quality are especially valued. Rather than drab and uniform products, people want flashier goods that are easily distinguishable.[65] Post-Fordist consumers are willing to pay more for distinctive, high-quality products.
- *Shorter production runs.* The more specialized products demanded in post-Fordist society require smaller and more productive systems. Huge factories producing uniform products are replaced by smaller plants turning out a wide range of products.
- *Flexible production.* New technologies make flexible production profitable. For example, computerized equipment that can be reprogrammed to produce different products replaces the old, single-function technology. This new production process is controlled through more flexible systems such as a more flexible form of management.

- *More capable workers.* Post-Fordist workers need more diverse skills and better training to handle the more demanding, more sophisticated technologies. These new technologies require workers who can handle more responsibility and operate with greater autonomy.
- *Greater differentiation.* As post-Fordist workers become more differentiated, they come to want more differentiated commodities, lifestyles, and cultural outlets. In other words, greater differentiation in the workplace leads to greater differentiation in the society as a whole. The result is more diverse demands from consumers and thus still greater differentiation in the workplace.

Although these elements of post-Fordism have emerged in the modern world, elements of old-style Fordism persist and show no signs of disappearing. There has been no clear historical break with Fordism; in fact, "McDonaldism," a phenomenon that clearly has many things in common with Fordism, is growing at an astounding pace in contemporary society. Among the things McDonaldism shares with Fordism are the following:

- Homogeneous products dominate a McDonaldized world. The Big Mac, the Egg McMuffin, and Chicken McNuggets are identical from one time and place to another.
- Technologies such as Burger King's conveyor system, as well as the French fry and soft drink machines throughout the fast-food industry, are as rigid as many of the technologies in Henry Ford's assembly-line system.
- The work routines in the fast-food restaurant are highly standardized. Even what the workers say to customers is routinized.
- The jobs in a fast-food restaurant are de-skilled; they take little or no skill.
- The workers are homogeneous and interchangeable.
- The demands and the actions of the customers are homogenized by the needs of the fast-food restaurant. Don't dare ask for a not-so-well-done burger; what is consumed and how it is consumed are homogenized by McDonaldization.

Fordism is thus alive and well in the modern world, although it has been transformed to a large extent into McDonaldism. Furthermore, classic Fordism—for example, in the form of the assembly line—remains a significant presence in American and world industry.

Postmodernity: A Threat to McDonaldization?

The more general theoretical perspective known as "postmodernism"[66] argues that we have entered, or are entering, a new society that represents a radical break with modern society; postmodernity follows and supplants modernity. Modern society is thought of as highly rational and rigid, whereas postmodern society is seen as less rational, more irrational, and more flexible. To the degree that postmodernity is seen as a successor to modernity, postmodern social theory stands in opposition to the McDonaldization thesis: The idea that irrationality is increasing contradicts the view that there is an increase in rationality. If we have, in fact, entered a new postmodern era, then McDonaldization would be confronted with a powerful opposing force.

However, less radical postmodern orientations allow us to see phenomena such as McDonald's as having both modern and postmodern characteristics.[67] David Harvey sees many continuities between modernity and postmodernity.[68] Central to Harvey's argument is the idea of time-space compression. He believes that modernism compresses both time and space, speeding the pace of life and shrinking the globe (for example, computers allowing us to send e-mail messages almost instantly to anywhere in the world), and that the process has accelerated in the postmodern era.

As an example of space compression within the McDonaldized world, consider that foods once available only in foreign countries or large cities are now quickly and widely available throughout the United States because of the spread of fast-food chains dispensing Italian, Mexican, or Cajun food. Similarly, in the realm of time compression, foods that formerly took hours to prepare can now take seconds in a microwave oven or be purchased in minutes at shops and supermarkets offering preprepared foods.

The best-known argument linking modernity and postmodernity is made by Fredric Jameson.[69] Jameson's position is that capitalism (certainly a "modern" phenomenon) is now in its "late" phase, although it continues to dominate today's world. The late phase of capitalism involves "a prodigious expansion of capital into hitherto uncommodified areas."[70] For Jameson, the key to contemporary capitalism is its multinational character and the fact that multinational corporations (such as McDonald's and IKEA) have greatly increased the range of products transformed into commodities.

Clearly, McDonaldization shows no signs of disappearing and being replaced by new, postmodern structures. However, McDonaldized systems do exhibit many postmodern characteristics side-by-side with modern elements (e.g., time, space compression). Thus, we are safe in saying that the McDonaldizing world demonstrates aspects of *both* modernity and postmodernity. And this conclusion clearly indicates that postmodernity does not represent a barrier to continued McDonaldization.

The Future: Are There Any Limits to the Expansion of McDonaldization?

The preceding section gives us a sense of where McDonaldization stands in the contemporary world and its relationship to some key ongoing changes. None of these changes are barriers to the further expansion of McDonaldization and may even serve to spur its further expansion. We turn now to an example—the climbing of Mt. Everest—which seems to indicate that there are few if any limits to the future expansion of McDonaldization.

The issue of the limits of McDonaldization is raised, at least implicitly, in Jon Krakauer's *Into Thin Air*, which describes several death-defying efforts to climb Mt. Everest in 1996. It is clear that there have been efforts to McDonaldize dangerous acts such as mountain climbing in general,[71] and Everest in particular, but it is also clear from the death of 12 people in the 1996 ascent that this "intrinsically irrational act"[72] cannot, at least as yet, be totally rationalized. Krakauer describes the variety of steps that have been taken over the years to McDonaldize the ascent.

At the top of the list are technological advances such as sophisticated climbing gear; canisters to supply supplemental oxygen at higher altitudes; helicopters for transport to the take-off point for the climb (a trip that used to take more than a month) and for the rescue of ill or injured climbers; medical technologies (and personnel) to deal with problems associated with falls, acute altitude sickness, and the like; and satellite phones, computers, the Internet, and fax machines to keep in touch with climbers as they ascend. Krakauer also cites organizational arrangements designed to prevent people from going it alone and to make the climbing team operate like a well-oiled machine. One group leader was lauded for his "impressive organizational skills" and his "elaborate systems."[73]

The best example of the effort to rationalize the climbing of Everest in 1996 was one team's system for "fast-track acclimatization" to the debilitating altitudes.[74] It involved fewer trips from base camp, and each successive excursion involved trekking a standard number of feet up the mountain. Overall, the "fast-track" method involved spending 4 weeks above 17,000 feet, 8 nights at 21,300 feet or higher, and only 1 night at 24,000 feet before climbers began the ascent to the 29,028-foot summit of Everest. The standard, less rationalized procedure involves climbers spending more time at 21,300 feet and higher, and there is at least one climb to 26,000 feet before attempting to reach the summit. On the basis of his observations, Krakauer cautiously concludes, "There is little question that extending the current eight—or nine—night acclimatization period at 21,000 to 24,000 feet would provide a greater margin of safety."[75]

Those who sought to McDonaldize the climbing of Everest can be described as trying to turn the route up the mountain into a nice, smooth, safe "toll road."[76] They stood to earn higher fees and to recruit more well-heeled climbers in the future if they could demonstrate that the dangers associated with climbing Everest were under control. Said one group leader, "We've got the big E figured out, we've got it totally wired. These days, I'm telling you, we've built a yellow brick road to the summit."[77]

The limitations of such efforts, however, are reflected in the argument that an expedition up Everest "can't be run like a Swiss train."[78] Several irrationalities of rationality are associated with such a McDonaldized effort. Traffic jams were created on the mountain because so many groups with so many people were attempting the climb. Paying climbers tended to be ill trained, ill prepared, and dependent on the guides. Most were strangers without much knowledge of one another's strengths and limitations as climbers. Because the climbers were paying large fees, leaders found it hard to say no to them even in situations where the climbers should have been refused.

But the best example of the irrationality of rationality relates to acclimatization. The fast-track method of acclimatization was efficient and allowed people to climb higher and faster and to get to the top quicker, but it depended on the use of bottled oxygen at higher levels. The lack of adequate time for acclimatization at each level made it harder for those involved in the 1996 climb to survive when oxygen proved unavailable at the higher levels.

The 1996 ascent exhibited other irrationalities, including inexperienced climbers, a guide who was climbing Everest for the first time, the apparent selfishness of another guide, the "damn-the-torpedoes" approach of one group leader, the competitive rivalry between group leaders, and the violation of one group's own deadline to begin descending from the summit. Such irrationalities are not inherent in the effort to climb Everest or in the rationalization of mountain climbing, but similar things are likely to occur and contribute to problems on any given climb.

Beyond the irrationalities of rationality and the irrationalities of this particular climb, there is the inherent irrationality of seeking to ascend Mt. Everest. Since organized climbs began in 1921, more than 200 people have died on Everest. As climbers ascend ever higher, physiological problems mount. Crevasses can shift, hurtling climbers to their death. Rocks careening down the mountain take their toll. The wind chill can reach −100°F. In fact, the biggest irrationality is the weather: In 1996, a completely unexpected storm killed those 12 climbers who were attempting to reach the peak, the largest single death toll in the history of Everest ascents.

The 1996 Everest disaster would seem to indicate that, despite people's best efforts, McDonaldization has limits. We certainly will never fully rationalize such death-defying activities as the climbing of Everest. But as we have seen, McDonaldization is not an all-or-nothing process; there are degrees of McDonaldization. Thus, many will continue to try to minimize the irrationalities associated with mountain climbing. In the case of Mt. Everest, future climbers can learn from past disasters and develop methods to minimize or eliminate the risks. The biggest danger is the unexpected storm, but improved weather-forecasting and -sensing devices can be developed and deployed.

While further accidents and deaths will mark future ascents of Everest (surprise storms will occur again), we may well be approaching something approximating a "yellow brick road to the summit" of Everest (perhaps operated by Disney). But by that time, it is likely that those daredevils who have found Everest so alluring will seek out some less-McDonaldized adventures or, better yet, find new ways of deMcDonaldizing the climbing of mountains such as Everest.

In fact, very recent developments in mountain climbing serve to make it less McDonaldized and more of an adventure. As the president of the American Alpine Club said, "Fifty years ago, adventure in the mountains was more about going places where no one had been. . . . Most of these places have been more thoroughly explored. Maybe adventure gets redefined."[79] One of the ways that it has been redefined is to begin eliminating the changes that served to McDonaldize climbing. For example, some climbers are choosing to be "free soloists" by refusing to use McDonaldizing protective gear such as ropes and harnesses to focus on the "simplicity and commitment of pure climbing."[80]

Others are seeking to deMcDonaldize climbing and make it more dangerous by trying to climb mountains as fast as they can. For example, the 13,000-foot Eiger mountain in Switzerland was first climbed in 1938. By 1950, it took 18 hours to climb the mountain. Most recently, the Eiger was climbed in 2 hours and 28 minutes. Greater speed makes it more difficult to McDonaldize the climb and therefore makes it more dangerous, more of an adventure. However, while this focus on increasing the speed of a climb makes climbing less McDonaldized, it is interesting to note that the emphasis is on one component of McDonaldization: calculability—specifically, increased speed. The emphasis on speed may well diminish the artistic aspects of climbing. In fact, one of the climbers focusing on speed said of himself, "It's not that I'm a great climber."[81]

There is great momentum to the process of McDonaldization, but natural limits and personal (to climb faster) interests also present some powerful barriers to it. While there may still be barriers in the natural and personal world, are there any to be found in the social and economic world? That is, is there anything to stop McDonaldization from altering everything about social and economic life not only in the United States but around the world? We will return to this question in Chapter 7, where we will see that there are barriers to McDonaldization. In fact, we will address in depth the possibility of deMcDonaldization.

3

Efficiency and Calculability

T his chapter deals with two of the four basic dimensions of McDonaldization: efficiency and calculability.

Efficiency: Drive-Throughs and Finger Foods

Efficiency is perhaps the dimension of McDonaldization most often linked to the seeming increase in the pace of contemporary life. Increasing efficiency is behind just-in-time production, faster service, streamlined operations, and tight schedules everywhere—from the workplace, to Disney World, to the home.

Efficiency is generally a good thing. It is clearly advantageous to consumers, who are able to obtain what they need more quickly with less effort. Similarly, efficient workers are able to perform their tasks more rapidly and easily. Managers and owners gain because more work gets done, more customers are served, and greater profits are earned. But as is the case with McDonaldization in general, and each of its dimensions, irrationalities such as surprising inefficiencies and the dehumanization of workers and customers emerge from the drive for increased efficiency.

Efficiency means choosing the optimum means to a given end. However, the truly optimum means to an end is rarely found. People and organizations rarely maximize because they are hampered by such factors as the constraints of history, financial circumstances, organizational realities and by the limitations of human nature.[1] Nevertheless, organizations continue to strive for maximization in the hope that they will at least increase their efficiency.

In a McDonaldized society, people rarely search for the best means to an end on their own; rather, they rely on previously discovered and institutionalized means. Thus, when people start a new job, they are not expected to figure out for themselves how to do the work most efficiently. Instead, they undergo training designed to teach them what has been discovered over time to be the most efficient way of doing the work. Once on the job, people may discover little tricks that help them perform the task more efficiently, and these days, they are encouraged to pass this information on to management so that all workers performing that task can perform a bit more efficiently. In this way, over time, efficiency (and productivity) gradually increases. In fact, much of the economic boom in the late 20th and early 21st centuries was attributed to dramatic increases in efficiency and productivity that permitted growth with little inflation. Even after the onset of the Great Recession beginning in late 2007, efficiency increased, but this time as employers discovered ways of producing as much, or more, with fewer and fewer employees.

Although the fast-food restaurant certainly did not create the yearning for efficiency, it has helped turn efficiency into an increasingly universal reality. Many sectors of society have had to change to operate in the efficient manner demanded by those accustomed to life in the drive-through lane of the fast-food restaurant. While many manifestations of efficiency can be traced directly to the influence of the fast-food restaurant, many more of them predate and helped shape the fast-food restaurant. Nonetheless, they all play a part in the preoccupation with efficiency fueled by McDonaldization.

Streamlining the Process

Above all else, Ray Kroc was impressed by the efficiency of the McDonald brothers' operation, as well as the enormous profit potential of such a system applied at a large number of restaurant sites. Here is how Kroc described his initial reactions to the McDonald's system: "I was fascinated by the simplicity and effectiveness of the system. . . . Each step in producing the limited menu was stripped down to its essence and accomplished with a minimum of effort. They sold hamburgers and cheeseburgers only. The burgers were all fried the same way."[2]

But Kroc's obsession with streamlined processes predated his discovery of McDonald's. When he was selling blenders to restaurants, he was disturbed by the restaurants' lack of efficiency: "There was inefficiency, waste, and temperamental cooks, sloppy service and food whose [sic] quality was never consistent. What was needed was a simple product that moved from start to completion in a *streamlined* path" (italics added).[3]

Kroc toyed with other alternatives for streamlining the restaurant meal before settling on the McDonald's hamburger as a model of efficiency: "He had contemplated hot dogs, then rejected the idea. There were too many kinds of hot dogs . . . there were all sorts of different ways of cooking hot dogs . . . boiled, broiled, rotisseried, charcoaled. . . . Hamburgers . . . were simplicity itself. The condiments were added to the hamburger, not built in. And there was only one way to prepare the hamburger—to grill it."[4]

Kroc and his associates experimented with each component of the hamburger to increase the efficiency of producing and serving it. For example, they started with partially sliced buns that arrived in cardboard boxes. But the griddle workers had to spend time opening the boxes, separating the buns, slicing them in half, and discarding the leftover paper and cardboard. Eventually, McDonald's found that buns sliced completely in half, separated, and shipped in reusable boxes could be used more efficiently. The meat patty received similar attention. For example, the paper between the patties had to have just the right amount of wax so that the patties would readily slide off the paper and onto the grill. Kroc's goal in these innovations was greater efficiency: "The purpose of all these refinements . . . was to make our griddle man's job easier to do quickly and well. And the other considerations of cost cutting, inventory control, and so forth were important to be sure, but they were secondary to . . . what happened there at the smoking griddle. This was the vital passage of our *assembly-line,* and the product had to flow through it smoothly or the whole plant would falter" (italics added).[5] To this day, efficiency remains the focus at McDonald's. For example, at a Washington, D.C., McDonald's, "the workers labored with an assembly-line efficiency."[6]

The Fast-Food Industry:
Speeding the Way From Secretion to Excretion

Today, all fast-food restaurants prepare their menu items on a kind of assembly line involving a number of people in specialized operations (for example, the burger "dresser"). The ultimate application of the assembly line to the fast-food process is, as pointed out in the preceding chapter, Burger King's conveyor belt: "Above and below the conveyor were two flames. Burgers were placed in this mesh, and they moved along the belt at a pre-set speed, were cooked on both sides simultaneously by the two flames, and then spilled out the other end into holding trays."[7]

Then, there is Domino's system:

Lonnie Lane starts slapping and saucing: kneading and tossing the dough and then spooning the proper measure of sauce on it. He slides the tray down . . . and Victor Luna starts reaching for the toppings. A dozen bins are arrayed in front of him: cheese, pepperoni, green pepper. . . . Luna sprinkles stuff over the tray by the handful. . . . He eases the tray onto a conveyor belt that takes it through a 12-foot oven in six minutes. . . . The store manager is dispatching waiting drivers and waiting drivers are folding pizza boxes. . . . The crew chief and quality controller . . . slices it with a pizza wheel and slides it into a box that already bears a computer label with the customer's address.[8]

Similar techniques are employed throughout the fast-food industry. To take one other example, many Krispy Kreme shops are "factory stores" where the donuts are produced using a highly efficient conveyor belt system. The donuts produced in this way are sold in the factory stores and shipped to Krispy Kreme outlets without such factories, as well as to supermarkets and other locales where the donuts are sold.

Getting diners into and out of the fast-food restaurant has also been streamlined. McDonald's has done "everything to speed the way from secretion to excretion."[9] Parking lots adjacent to the restaurant offer readily available parking spots. It's a short walk to the counter, and although customers sometimes have to wait in line, they can usually quickly order, obtain, and pay for their food. The highly limited menu makes the diner's choice easy, in contrast to the many choices available in other restaurants. ("Satellite" and "express" locations are even more streamlined.) With the food obtained, it is but a few steps to a table and the beginning of the "dining experience." With little inducement to linger, diners generally eat quickly and then gather the leftover paper, Styrofoam, and plastic, discarding them in a nearby trash receptacle, and get back in their cars to drive to the next (often McDonaldized) activity.

Those in charge of fast-food restaurants discovered that the drive-through window made this whole process far more efficient. Instead of requiring diners to undergo the "laborious" and "inefficient" process of parking the car, walking to the counter, waiting in line, ordering, paying, carrying the food to the table, eating, and disposing of the remnants, the drive-through window offered diners the streamlined option of driving to the window and driving off with the meal. Diners could eat while driving if they wanted to be even more efficient. The drive-through window is also efficient for the fast-food restaurant. As more and more people use the drive-through window, fewer parking spaces, tables, and employees are needed. Furthermore, consumers take their debris with them as they drive away, thereby reducing the need for trash receptacles and employees to empty those receptacles periodically.

Modern technology offers further advances in streamlining. Here is a description of some of the increased efficiency at a Taco Bell in California:

> Inside, diners in a hurry for tacos and burritos can punch up their own orders on a touch-screen computer. Outside, drive-through customers see a video monitor flash back a list of their orders to avoid mistakes. They then can pay using a pneumatic-tube like those many banks employ for drive-up transactions. Their food, and their change, is waiting for them when they pull forward to the pickup window. And if the line of cars grows too long, a Taco Bell worker will wade in with a wireless keyboard to take orders.[10]

Similarly, McDonald's is planning to use such touch screens in Europe and to use them to streamline the process of ordering and paying for food by eliminating counter people and cashiers.[11]

To further increase efficiency, an increasing number of fast-food restaurants are accepting credit and debit cards.

Home Cooking (and Related Phenomena):
"I Don't Have Time to Cook"

In the early 1950s, the dawn of the era of the fast-food restaurant, the major alternative to fast food was the home-cooked meal, made mostly from ingredients purchased beforehand at various local stores and early supermarkets. This was clearly a more efficient way of preparing meals than earlier methods, such as hunting game and gathering fruits and vegetables. Cookbooks also made a major contribution to efficient home cooking. Instead of inventing a dish every time a meal was prepared, the cook could follow a recipe and thus more efficiently produce the dish.

Soon, the widespread availability of the home freezer led to the expanded production of frozen foods. The most efficient frozen food was (and for some still is) the "TV dinner." Swanson created its first TV dinner, its meal-in-a-box, in 1953 and sold 25 million of them in the first year.[12] The large freezer also permitted other efficiencies, such as making a few trips to the market for enormous purchases rather than making many trips for small ones.

Meals from the freezer began to seem comparatively inefficient, however, with the advent of microwavable meals.[13] Microwaves usually cook faster than other ovens, and people can prepare a wider array of foods in them. Perhaps most important, microwave ovens spawned a number of food products (including microwavable soup, pizza, hamburgers, fried chicken, French fries, and popcorn) reminiscent of the fare people have learned to love in fast-food restaurants. For example, one of the first microwavable foods produced by Hormel was an array of biscuit-based breakfast sandwiches popularized by McDonald's with its Egg McMuffin.[14] As one executive put it, "Instead of having a breakfast sandwich at McDonald's, you can pick one up from the freezer of your grocery store."[15] In some ways, "homemade" fast foods seem more efficient than the versions offered by fast-food restaurants. Instead of getting into the car, driving to the restaurant, and returning home, people need only pop their favorite foods into the microwave. However, the microwaved meal requires a prior trip to the market.

Supermarkets have long been loaded with other kinds of products that increase efficiency for those who want to "cook" at home. Instead of starting from scratch, the cook can use prepackaged mixes to make "homemade" cakes, pies, pancakes, waffles, and many other foods. In fact, entire meals are now available right out of the box. Dinty Moore's Classic Bakes are entire casserole dinners (e.g., beef stew and biscuits) for four to five people and promise to be "hot and hearty, quick and convenient, ready in minutes."

An increasingly important competitor is the fully cooked meal consumers may now buy at the supermarket. People can merely stop on the way home and purchase prepackaged main courses and even entire meals, which they "prepare" by unwrapping the packages—no cooking required.

The McDonaldization of food preparation and consumption also encompasses the booming diet industry. Losing weight is normally difficult and time-consuming, but diet books promise to make it easier and quicker. The preparation of low-calorie food has also been streamlined. Instead of cooking diet foods from scratch, dieters may now purchase an array of prepared foods in frozen or microwavable form. Those who do not wish to go through the inefficient process of eating these diet meals can consume products such as diet shakes and bars (Slim-Fast, for example) in a matter of seconds. Dieters seeking even greater efficiency have access to various pills that expedite weight loss—the now-banned "fen-phen" and the still available Xenical. Then there is cosmetic surgery and even more radical and invasive steps such as gastric bypass surgery.

The issue of dieting points to new efficiencies outside the home as well, that is, to the growth of diet centers such as Jenny Craig, NutriSystem,[16] Weight Watchers, and Curves (really an exercise center with a highly efficient 30-minute workout).[17] (There are now even a large number of apps available on smartphones dealing with weight loss and health that can be accessed easily and that make weight loss even more efficient.) NutriSystem sells dieters, at substantial cost, prepackaged freeze-dried food.

Shopping: Creating
Ever-More Efficient Selling Machines

Shopping for all kinds of goods and services, not just food, has also been streamlined. The department store obviously is a more efficient place in which to shop than a series of specialty shops dispersed throughout the city or suburbs. The shopping mall increases efficiency by bringing a wide range of department stores and specialty shops under one roof.[18] It is cost-efficient for retailers because the collection of shops and department stores brings in throngs of people ("mall synergy"). And it is efficient for consumers because, in one stop, they can visit numerous shops, have lunch at a "food court" (likely populated by many fast-food chains), see a movie, have a drink, and go to an exercise or diet center.

The drive for shopping efficiency did not end with the malls. 7-Eleven and its clones (for example, Circle K, AM/PM, and Wawa) have become drive-up, if not drive-through, minimarkets. For those who need only a few items, pulling up to a highly streamlined 7-Eleven (more than 42,000 locations worldwide[19]) is far more efficient (albeit more costly) than running into a supermarket.

Shoppers have no need to park in a large lot, obtain a cart, wheel through myriad aisles in search of needed items, wait in lines at the checkout, and then tote purchases back to a sometimes distant car. At 7-Eleven, they can park right in front and quickly find what they need. Like the fast-food restaurant, which offers a highly circumscribed menu, 7-Eleven has sought to fill its shops with a limited array of commonly sought goods: bread, milk, cigarettes, aspirin, even videos, and self-serve items such as hot coffee, hot dogs, microwaved sandwiches, cold soda, and Slurpees. 7-Eleven's efficiency stems from the fact that it ordinarily sells only one brand of each item, with many items unobtainable.

Even more efficient are the quick and convenient BrewThrus, a small chain based primarily in North Carolina. Customers simply drive into a BrewThru, which is set up like a garage lined on both sides with many convenience store products, especially beer (including kegs) and wine. An attendant comes out to the car to take your order, brings you what you want, takes your money, and you are back on the road in no time.[20]

For greater selection, consumers must go to the relatively inefficient supermarket. Of course, supermarkets have sought to make shopping more efficient by institutionalizing automated self-checkout lanes (see below) and 15-item-limit, no-checks-accepted lines for consumers who might otherwise frequent the convenience stores.

People who do not feel that they have the time to visit the mall are able to shop from the comfort of their homes through catalogs (for example, L.L. Bean or Lands' End) made more efficient these days by being available online.[21] Another alternative to visiting the mall is home television shopping. The efficiency of shopping via catalogs and TV is increased even further with express package delivery systems, such as Federal Express and UPS.

The Internet is the most important factor in greatly increased shopping efficiency. For example, instead of traveling to a book superstore (dying as evidenced by the 2011 bankruptcy and closing of all Borders superstores and the difficulties being experienced by Barnes & Noble) or wandering from one small bookshop (also dying) to another, you can access Amazon.com and millions of different titles at your fingertips.[22] After selecting and charging the titles you want, you just sit back and wait for the books to be delivered to your door, usually within a day or two. The advent of e-books and e-readers such as Amazon's Kindle have made shopping for books infinitely more efficient since the books are now ordered and downloaded almost instantaneously. In fact, for the first time, Amazon sold more books via Kindle in 2011 than it did both hardcover and paperback print books.[23]

Of course, there are innumerable other Internet sites (e.g., overstock.com) where one can efficiently shop for virtually anything. Then there is eBay .com, which allows buyers and sellers to deal with one another in a highly efficient manner. (We will have a lot more to say about eBay from the point of view of McDonaldization *and* deMcDonaldization in Chapter 7.) "Virtual pharmacies" allow people to obtain prescription drugs without seeing a physician; consultations with "online doctors" are also available via, for example, HealthNation.[24]

An often overlooked aspect of the efficiency of cybershopping is that it can be done while you are at work.[25] Although employers are likely to feel that shopping from work adversely affects worker efficiency, it is certainly very efficient from the perspective of the worker/consumer.

Of course, the drive to make Internet shopping ever more efficient continues. There are now shopping robots, or "online comparison services," that automatically surf the Web for specific products, lowest prices, and shortest delivery dates.[26] For example, Google.com has "Google Product Search": "Browse by category—apparel, computers, flowers, whatever—or enter a query term, and it will present a list of matching products, each with a thumbnail sketch on the left and description, price and retailer on the right."[27] However, it has now become even more efficient just to go to Amazon.com, which has virtually everything one might want at competitive prices and without the need for online comparison price services.

All types of shopping, but particularly ordering from distant merchants, have become far more efficient with the widespread use of credit cards.

Higher Education: Just Fill in the Box

In the contemporary university (now often dubbed "McUniversity"),[28] efficiency is to be found—for example, the machine-graded, multiple-choice examination. In a much earlier era, students were examined individually in conference with their professors. Later, the essay examination became popular. Grading a set of essays was more efficient than giving individual oral examinations, but it was still relatively time-consuming. Enter the multiple-choice examination, the grading of which was a snap. In fact, graduate assistants could grade it, making evaluation of students even more efficient for the professor. Computer-graded examinations maximize efficiency for both professors and graduate assistants.

The computerized Blackboard system has added still greater efficiencies. For example, it eliminates the need for professors to reproduce and distribute materials to class. It also makes it possible to take exams online rather than in class. Furthermore, Blackboard grades the exams, adds them to the grade book, and even calculates a final grade.

Other innovations in academia have further streamlined the educational process. Instead of professors composing multiple-choice questions, publishers provide digital, online sets of questions. Another advance is computer-based programs to grade essay examinations (and even term papers).[29] Professors can now choose to have very little to do with the entire examination process, from composing questions to grading, freeing up time for activities that many professors, but few students, value more highly, such as writing and research.

Publishers have provided other services to streamline teaching for those professors who adopt best-selling textbooks, including lecture outlines, Power-Points, text-related websites, computer simulations, discussion questions, DVDs, movies, and ideas for guest lecturers and student projects. Professors who choose to use all these devices need do little or nothing on their own for their classes. Students are also increasingly able to be more efficient *in* class by using their laptops and smartphones to multitask in various ways. This can be educationally beneficial when it involves doing Google searches during a lecture, but it can have adverse effects when students are playing games online, writing on someone's Facebook wall, or texting on their smartphones. Also worth noting is ratemyprofessors.com, where students can efficiently evaluate their professors as well as find ratings by other students.

Wikipedia has become an efficient source of information for both professors and students. There was a time when most professors were critical of the use of Wikipedia, but more and more have come to embrace it. They use it themselves and allow, even encourage, their students to use it.

One other academic efficiency worth noting is the ability of students to purchase already completed term papers online. A variety of websites[30] now promise to deliver original, customized research papers on any topic for a "low, low fee" of, say, $8.99 per page. You could (although it is not advised) purchase a 3-plus page paper on McDonaldization on one website for $19.95. Websites even have quick service and express delivery available ($14.99 per page if you need the paper in 48 hours) for those students who have put off academic dishonesty to the last moment. Beware, however, for there are also other websites (e.g., turnitin.com) that help professors detect plagiarism, thereby combating student gains in efficiency through plagiarism with an efficient system to detect it.[31]

Health Care: Docs-in-a-Box

It might be assumed that modern medicine is immune to the drive for efficiency and invulnerable to rationalization more generally.[32] However, medicine has been McDonaldized. One example is Dr. Denton Cooley (his "fetish is efficiency"), who gained worldwide fame for streamlining delicate open-heart surgery in a "heart surgery factory" that operated "with the precision of an assembly-line."[33] Even more striking is the following description of the Moscow Research Institute of Eye Microsurgery, which promises on its website "efficient therapy of most serious ophthalmic Diseases"[34]:

In many ways the scene resembles any modern factory. A conveyor glides silently past five work stations, periodically stopping, then starting again. Each station is staffed by an attendant in a sterile mask and smock. The workers have just three minutes to complete their tasks before the conveyor moves on; they turn out 20 finished pieces in an hour.

Nearly everything else about the assembly line, however, is highly unusual: the workers are eye surgeons, and the conveyor carries human beings on stretchers. This is where the production methods of Henry Ford are applied to the practice of medicine . . . a "medical factory for the production of people with good eyesight."[35]

Such assembly lines are not yet the norm in medicine, but one can imagine that they will grow increasingly common in the coming years.

While not quite done on an assembly line, I had laser eye surgery in 2011 done on the basis of a system that is very close to it. Ten patients requiring the same operation were told to report at a given hour, ushered into a room, and told to sit in a specific place in a line of chairs. On several occasions, assistants went down the line putting various drops into the eye that was to be operated on. When the surgeon arrived, each patient was hustled—in order of his or her position in the line of chairs—into another room, told by another assistant to place his or her chin on a chin rest on the laser machine, and soon heard a series of zapping sounds. When the sound stopped and the surgery was completed, the patients were dismissed and ushered back to their original seats. When all 10 patients had their surgery, each in turn was brought back to the room where the surgery was performed, although by this time the surgeon was long gone. Other assistants checked the eye to be sure the surgery was successful, and if it was, the patient was told he or she could leave.

Also increasing is the use of robots to perform advanced forms of surgery. Perhaps the best known is the DaVinci system that is revolutionizing various forms of surgery (e.g., for prostate cancer). This minimally invasive system not only makes an operation more efficient but also makes the process more efficient from a patient's point of view. Because only small incisions are made, hospital stays are reduced to perhaps a day, and postoperative recovery time is relatively brief.[36]

Perhaps the best example of the increasing efficiency of medical practice in the United States is the growth of walk-in/walk-out surgical or emergency centers such as DR (Duane Reade) Walk-In Medical Care clinics in New York City.[37] "McDoctors," or "Docs-in-a-box," serve patients who want highly efficient medical care. Each center handles only a limited number of minor problems but with great dispatch. Although stitching a patient with a minor laceration cannot be done as efficiently as serving a customer a hamburger, many of the same principles apply. For instance, it is more efficient for the patient to walk in without an appointment than to make an appointment with a regular physician and wait until that time arrives. For a minor emergency, such as a slight laceration, walking through a McDoctors is more efficient than working your way through a large hospital's emergency room. Hospitals are set up to handle serious problems for which efficiency is not (yet) the norm, although some hospitals already employ specialized emergency room physicians and teams of medical personnel. Docs-in-a-box are also more efficient than private doctors' offices because they are not structured to permit the kind of personal (and therefore inefficient) attention patients expect from their private physicians.

A more recent development is minute clinics found in pharmacies (e.g., CVS) and even in supermarkets, discounters, and big-box stores.[38] These may be staffed by nurse practitioners and physician assistants and offer efficient help in the case of minor medical matters. It has become increasingly common to get flu shots in a neighborhood pharmacy or supermarket (perhaps offered near the meat department and by a butcher—just kidding!).

Entertainment: Moving People (and Trash) Efficiently

Many people no longer deem it efficient to trek to their local theater to see a movie. For a time, DVDs, and the stores that rented them, boomed. Blockbuster, at one time the largest video rental firm in the United States, considered "itself the McDonald's of the video business."[39] However, Blockbuster went bankrupt in late 2010.

The video rental business was transformed by more efficient alternatives. Blockbuster Express now has thousands of video rental machines (Redbox—once owned by McDonald's—is another major player in this area). However, this pales in comparison to the streaming of movies from companies such as Netflix (Blockbuster has a similar system, as do Amazon.com and iTunes) or even streaming them free, either legally or illegally. Then there are the pay-per-view movies offered by many cable companies. And there is Hulu.com, which allows people to download, for a fee, any of its catalogue of current primetime TV shows, classic TV programs, movies, documentaries, and so on.[40] These can be viewed not only at home but on a variety of mobile systems. DVRs and TIVO permit customers to record their favorite shows while they are watching something else or to rewind or pause live television. For those not satisfied with just a single offering, there is "picture-in-a-picture," which enables them to view a movie while also watching a favorite TV show on an inset on the screen. Then there are the satellite networks that allow you to watch several football games at once; with the DISH Network's "Red Zone," one is able to watch all games when the ball is within the opponent's 20-yard line.

Another sort of efficiency in the entertainment world is the system for moving people developed by modern amusement parks, particularly Disneyland and Walt Disney World.[41] A system of roads filters thousands of cars each day into the appropriate parking lots. Jitneys whisk visitors to the gates of the park. Once in the park, they find themselves in a vast line of people on what is, in effect, a huge conveyor belt that leads them from one ride or attraction to another. Once they actually reach an attraction, a conveyance such as a car, boat, submarine, plane, rocket, or moving walkway moves them rapidly through and out of the attraction.

Disney World has been victimized by its own success: Even its highly efficient systems cannot handle the hordes that descend on the park at the height of the tourist season. Since 1999, Disney has sought to deal with this problem with its FASTPASS system that allows a visitor to arrange a specific time to be at a given attraction and to enter via a separate and much faster-moving FASTPASS line. Of course, there are limits on the number of FASTPASSes that can be issued. It would be self-defeating if every visitor used a FASTPASS for every trip to every attraction. There are still long lines at Disney resorts, and even the FASTPASS lines may not be so fast.

People are not the only things Disney World must process efficiently.[42] The throngs that frequent such amusement parks eat a great deal (mostly fast food, especially finger foods) and therefore generate an enormous amount of trash. If Disney World emptied trash receptacles only at the end of each day, the barrels would be overflowing most of the time. To prevent this eyesore (and it must be prevented since order and cleanliness—some would say sterility—are key components of the McDonaldized world in general and Disney World in particular), hordes of employees constantly sweep, collect, and empty trash. To take a specific example, bringing up the rear in the nightly Disney parade is a group of cleaners who almost instantly dispose of whatever trash and animal droppings have been left behind. Within a few minutes, they have eliminated virtually all signs that a parade has just passed by. Disney World also employs an elaborate system of underground tubes. Garbage receptacles are emptied into this system, which whisks the trash away at about 60 miles per hour to a central trash-disposal plant far from the view of visitors. Disney World is a "magic kingdom" in more ways than one. Here is the way one observer compares another of the modern, highly rational amusement parks—Busch Gardens—to ancestors such as county fairs and Coney Island: "Gone is the dusty midway. . . . In its place is a vast, self-contained environment . . . endowed with the kind of *efficiency* beyond the reach of most cities" (italics added).[43]

Online and Smartphone Dating:
Show Your Interest With Just a "Wink"

For young people, dating has become "dated," inefficient in an era in which they can simply "hang out" together. However, hanging out becomes less possible as people grow a bit older. Dating is a highly inefficient process that has been streamlined as a result of the Internet and the ability to find and make dates on online services such as eHarmony.com and match.com.[44] (For those who are more interested in sex than the less efficient process of dating, there are sites such adultfriendfinder.com and Craigslist Casual Encounters.[45]) With a single click, one can find men or women within a specified age group and a given distance from one's zip code (no long, unnecessary trips needed). Key words are provided on the site (e.g., "charming," "energetic"), making it easier to find a particular kind of person. It is possible to scroll quickly through hundreds of profiles of potential dates who meet given criteria. Once a profile of interest is located, a simple click indicates a "wink" at a potential date. Other clicks can organize potential dates into a "favorites" list so that, if one possibility does not pan out, another can be located quickly and contacted.

If a date is chosen, it is likely that sooner or later, the relationship will not work out. Once that happens, it is possible to block the spurned suitor's access to one's profile. Best of all, a person can be back on the dating scene in an instant with a plethora of alternatives on the website or on one's favorites list.

Even more recent and efficient is the use of smartphones and mobile dating services, relevant apps, and geo-locating technology to find a date. This can be done on the go; one no longer needs to be tied to a computer. As a result, dates can be arranged even more quickly and efficiently.

Other Settings: Ten Steps to Spiritual Maturity

Modern health clubs, including chains such as Bally Total Fitness, Gold's Gym, and 24-Hour Fitness, also strongly emphasize efficiency.[46] These clubs often offer virtually everything needed to lose weight and stay in shape, including exercise machines, a running track, and a swimming pool. The exercise machines are highly specialized so that people can efficiently work specific areas of the body. Running machines, StairMasters, elliptical runners, and recumbent spinnakers increase cardiovascular fitness; weight-lifting machines increase strength and muscularity in targeted areas of the body. Many machines keep track of exactly how many calories are being burned away. People can watch TV, read, or listen to music while working out.[47]

Other examples of streamlining to increase efficiency abound. Drive-through windows and ATMs streamline banking for both consumers and bankers. Smartphones allow people to snap pictures and send them instantaneously to others with similar phones (or to e-mail accounts). At gasoline stations, customers put their credit cards into a slot and have no contact with, or work done by, anyone working for the gas station. In the case of ExxonMobil's "Speedpass," a transponder attached to a key tag or to a vehicle's rear window communicates with the pump via radio frequency signals (a similar technology is now used on most of the nation's toll roads). When the car pulls up, the pump is activated, and the correct amount is charged to the driver's credit card account. Of growing importance is the ability to pay for things via the cell phone.

Even religion has been streamlined, with drive-in churches and televised religious programs.[48] Christian bookstores are stocked with "how-to" books "claiming to be able to teach us the 10 steps to 'spiritual maturity' or how to be a successful parent in 60 minutes."[49]

Search engines such as Google, Yahoo!, and Bing now do a lot of the work that was formerly performed by computer users.[50] In the early days of the Internet, getting the information one wanted was a difficult matter requiring a great deal of skill and knowledge of arcane computer programs. Now, all users need do is access the search engine and type in the desired topic, and they are on their way. A process of de-skilling has taken place; users no longer need to have many skills—they are now built into the system. More recently, folksonomies such as delicious, Flickr, YouTube, and digg enable people to search efficiently using other users' tags. Tagging is a process whereby users attach short (usually one-word) descriptions to things like photos on Flickr. If you wanted to find a Flickr picture of a clown riding a bike, you could type "clown" and "bike," and all the photographs with those two "tags" would be displayed.

The Internet also renders such activities as political campaigning[51] and medical symposia[52] more efficient. Even more obviously, e-mail is far more streamlined than "snail mail."[53] Instead of writing a letter, all that is needed are a few keystrokes (one is limited to 140 of them on Twitter). Better yet, if both you and the person you are e-mailing are simultaneously using the same Internet chat service (for example, Google Talk known as "Gchat"), you can have a "conversation" online without any need to send even an e-mail. A similar set of advantages has attracted people to e-cards for birthdays, anniversaries, holidays, and the like.[54]

Texting is not only an efficient method of communication but has brought with it a great streamlining of speech among and between people. As a result, ever shorter ways of saying things are emerging, including such common shorthand as ur = you are, l8r = later, and emoticons such as :) = smile and ;) = wink.

Simplifying the Product

Another way to increase efficiency is by simplifying products. For example, complex foods based on sophisticated recipes are, needless to say, not the norm at fast-food restaurants. The staples of the industry are foods (e.g., hamburgers, slice of pizza) that require relatively few ingredients and are simple to prepare, serve, and eat. In fact, fast-food restaurants generally serve "finger food," food that can be eaten without utensils.

Many innovations over the years have greatly increased the number and types of finger foods available. The Egg McMuffin is an entire breakfast—egg, Canadian bacon, English muffin—combined into a handy sandwich. Devouring such a sandwich is far more efficient than sitting down with knife and fork and eating a plate full of eggs, bacon, and toast. The creation of the Chicken McNugget, perhaps the ultimate finger food, reflects the fact that chicken is pretty inefficient as far as McDonald's is concerned. The bones, gristle, and skin that are such a barrier to the efficient consumption of chicken have all been eliminated in the Chicken McNugget. Customers can pop the bite-sized morsels of fried chicken into their mouths even as they drive. Were they able to, the mass purveyors of chicken would breed a more efficiently consumed chicken free of bones, gristle, and skin.[55] Because it is completely encased in dough, McDonald's apple pie can be munched like a sandwich.

McDonald's continues to experiment with new menu items. One that has been successful is the snack wrap.[56] This is a classic McDonaldized food. For one thing, it is another form of "finger food" and can be eaten quickly and efficiently. For another, it uses ingredients already in McDonald's restaurants and used in other menu items—breaded chicken strips, flour tortillas, shredded lettuce and cheese, and ranch sauce.

The limited number of menu choices also contributes to efficiency in fast-food restaurants. McDonald's does not serve egg rolls, and Taco Bell does not offer fried chicken. Advertisements like "We do it your way" or "Your way, right away" imply that fast-food chains happily accommodate special requests. But pity the consumer who has a special request in a fast-food restaurant. Because much of their efficiency stems from the fact that they virtually always do it one way—their way—the last thing fast-food restaurants want is to do it your way. Customers with the temerity to ask for a less well-done burger or well-browned fries are likely to cool their heels for a long time waiting for such "exotica." Few customers are willing to wait because, after all, it undermines the main advantages of going to a fast-food restaurant. The idea behind what Henry Ford said long ago about cars has been extended to hamburgers: "Any customer can have a car painted any color that he wants so long as it is black."[57]

Many products other than fast food have been simplified in the name of efficiency. AAMCO Transmissions works mainly on transmissions, and Midas Muffler largely restricts itself to the installation of mufflers. H&R Block does millions of simple tax returns in its nearly 13,000 offices. Because it uses many part-time and seasonal employees and does not offer the full array of tax and financial services available from a CPA, it is undoubtedly not the best place to have complicated tax returns completed.[58] "McDentists" may be relied on for simple dental procedures, but people would be ill advised to have root canal work done there. Pearle Vision centers offer eye examinations, but people should go to an eye doctor for any major vision problem. *USA TODAY* offers readers highly simplified "News McNuggets."

Putting Customers to Work

A final mechanism for increasing efficiency in a McDonaldizing world is to put customers to work. (As will be discussed in Chapter 7, consumers who also produce [work] are called "prosumers."[59]) Fast-food customers perform many more unpaid tasks than those who dine at full-service restaurants: "McDonald's came up with the slogan 'We do it all for you.' In reality, at McDonald's, we [the customers] do it all for them. We stand in line, take the food to the table, dispose of the waste, and stack our trays. As labor costs rise and technology develops, the consumer often does more and more of the work."[60] While it is efficient for the fast-food restaurant to have consumers wait in line, waiting in line is inefficient for consumers. Is it efficient to order your own food rather than having a waiter do it? Or to bus your own paper, plastic, and Styrofoam rather than having a busperson do it?

The tendency to put customers to work was underscored by Steak 'n Shake's (nearly 500 restaurants in the United States[61]) TV advertisements describing fast-food restaurants as "workaurants."[62] In contrast, Steak 'n Shake emphasizes its use of china plates and the fact that the food is actually served by a wait staff.

The salad bar is a classic example of putting the consumer to work. The customer "buys" an empty plate and then loads up on the array of vegetables and other foods available that day. Quickly seeing the merit in this system, many supermarkets installed their own, more elaborate salad bars. The salad lover can thus work as a salad chef at lunch hour in the fast-food restaurant and then do it all over again in the evening at the supermarket. The fast-food restaurant and the supermarket achieve huge gains in efficiency because they need only a small number of employees to keep the various compartments well stocked.

There is an all-you-can-eat restaurant chain, Souplantation (called Sweet Tomatoes in some areas), with 112 outlets throughout the United States.[63] Its main attraction is a lengthy salad bar (really a kind of assembly line) that customers encounter as they enter the restaurant. At lunch and dinner times, there are often long lines on both sides of the salad bar. In fact, at particularly busy times, the lines snake out the door and into the parking lot. As diners move slowly along the salad bar, they fill their plates with the desired foods. At the end of the salad bar are two cash registers where those in each line pay for their food. Various other foods and desserts are available at counters in the restaurant, and after they have finished their salads, customers trek to them, sometimes over and over again.[64]

In a number of fast-food restaurants, including Fuddruckers (140 franchises)[65] and Roy Rogers,[66] consumers are expected to take a naked burger to the "fixin' bar" to add lettuce, tomatoes, onions, and so on. The customers thus end up logging a few minutes a week as sandwich makers. At Burger King and most other fast-food franchises, people must fill their own cups with ice and soft drinks, thereby spending a few moments as "soda jerks." Similarly, customers serve themselves in the popular breakfast buffets at Shoney's or the lunch buffets at Pizza Hut.

As pointed out earlier, McDonald's is testing self-ordering kiosks in some restaurants that allow customers to use a touch screen to place their food orders. They do what counter people at McDonald's currently do—find and touch the picture on the screen that matches the food being ordered.[67] Also of note is the trend toward self-ordering food kiosks at gas station travel plazas, such as Sheetz and Wawa. One company that makes these kiosks for a wide variety of restaurants at airports, drive-throughs, and casinos is Nextep. It claims that these systems outperform humans by displaying attractive pictures of food and offering extras to entice the customer to purchase more food.[68]

Shopping also offers many examples of imposing work on the consumer. The old-time grocery store, where the clerk retrieved the needed items, has been replaced by the supermarket, where a shopper may put in several hours a week "working" as a grocery clerk, seeking out wanted (and unwanted) items during lengthy treks down seemingly endless aisles. Having obtained the groceries, the shopper then unloads the food at the checkout counter and, in some cases, even bags the groceries.

Of course, some supermarket checkout stands now ask the customer to do the scanning, thereby eliminating the need for a checkout clerk.[69] Such self-checkout terminals are increasing rapidly in the United States and throughout the world, and the number is expected to quadruple by 2014.[70] The systems that allow customers to pay with credit cards eliminate the need for cashiers. The developer of one scanning system predicted that soon "self-service grocery technology could be as pervasive as the automatic cash machines used by bank customers."[71] One customer, apparently a strong believer in McDonaldization, said of such a system, "It's quick, easy and efficient. . . . You get in and out in a hurry."[72] But as an official with a union representing supermarket clerks put it, "To say it's more convenient for the customer is turning the world upside down. . . . In general, making customers do the work for themselves is not customer service."[73]

Virtually gone are gas station attendants who fill gas tanks, check the oil, and clean windows; people now put in a few minutes a week as unpaid attendants. Although one might think that eliminating the gas station attendant leads to lower gasoline prices (and indeed it does in the short run), a comparison of gas prices at stations with and without attendants shows little difference in price. In the end, the gasoline companies and service station owners simply found another way to force the consumer to do the work employees once had to be paid to perform.

In some doctors' offices, patients must now weigh themselves and take their own temperatures. More important, patients have been put to work in the medical world through the use of an increasingly wide array of do-it-yourself medical tests. Two basic types are available: monitoring instruments and diagnostic devices.[74] Monitoring devices include blood pressure monitors and glucose and cholesterol meters, as well as at-home kits marketed to parents to allow them to test their children for marijuana, heroin, and so on. Among the diagnostic tests are pregnancy detectors, ovulation predictors, HIV test kits, and fecal occult blood detectors. Thus, patients are now being asked to familiarize themselves with technologies that were formerly the exclusive province of physicians, nurses, or trained technicians. In addition, patients are being asked to provide samples of bodily fluids (blood, urine) or by-products (fecal matter) that were once handled (very carefully) by professional medical people. But in an era of high medical costs, it is cheaper and more efficient (no unnecessary trips to the doctor's office or to the lab) for patients to monitor and test themselves. Such home testing may identify problems that otherwise might not be discovered, but it can also lead to unnecessary worry, especially in the case of "false-positive" results. In either case, many of us are now "working," at least part-time, as unpaid medical technicians.

The automated teller machine (ATM) in the banking industry allows everyone to work, for at least a few moments, as an unpaid bank teller (and often pay fees for the privilege). To encourage the use of ATMs, some banks have begun charging a fee for using human tellers.[75]

When a receiver fails, DISH Network mails its customers a new one as a replacement. The customer is expected to return the defective one in the same box that contained the new one. More important, it is up to the consumer to install the new receiver. There is, of course, help available by phone if necessary—and it is necessary! DISH will send someone to install the new receiver, but the time delay involved and the relatively high cost discourage most customers from taking this option.

When calling many businesses these days, instead of dealing with a human operator, people must push a bewildering sequence of numbers and codes before they get, they hope, the desired extension.[76] Here is the way one humorist describes such a "conversation" and the work involved for the caller:

The party you are trying to reach—Thomas Watson—is unavailable at this time. To leave a message, please wait for the beep. To review your message, press 7. To change your message after reviewing it, press 4. To add to your message, press 5. To reach another party, press the star sign and enter the four-digit extension. To listen to Muzak, press 23. To transfer out of phone mail in what I promise you will be a futile effort to reach a human, press 0—because we treat you like one.[77]

Instead of being interviewed by the government census taker, people usually receive a questionnaire (one that is supposedly self-explanatory) in the mail to fill out on their own. The self-response rate for occupied housing units was 74% in the 2010 census.[78] Census takers were used about a quarter of the time to obtain the information, and even then they were deployed only after residents failed to respond to the mailed questionnaire.[79]

The ubiquity of all of these seemingly trivial activities means that the modern consumer spends an increasingly significant amount of time and energy doing unpaid labor. Although organizations are realizing greater efficiencies, customers are thus often sacrificing convenience and efficiency.

Calculability: Big Macs and Little Chips

McDonaldization also involves calculability: calculating, counting, quantifying many different things. Quantity tends to become a surrogate for quality.[80] Numerical standards are set for both processes (production, for example) and end results (goods, for example). In terms of processes, the emphasis is on speed (usually high), whereas for end results, the focus is on the number of products produced and served or on their size (usually large).

This calculability has a number of positive consequences, the most important being the ability to produce and obtain large amounts of things very rapidly. Customers in fast-food restaurants get a lot of food quickly; managers and owners get a great deal of work from their employees, and the work is done speedily. However, the emphasis on quantity tends to affect adversely the quality of both the process and the result. For customers, calculability often means eating on the run (hardly a "quality" dining experience) and consuming food that is almost always mediocre. For employees, calculability often means obtaining little or no personal meaning from their work; therefore, the work, products, and services suffer.

Calculability is intertwined with the other dimensions of McDonaldization. For instance, calculability makes it easier to determine efficiency; that is, those steps that take the least time are usually considered the most efficient. Once quantified, products and processes become more predictable because the same amounts of materials or time are used from one place or time period to another.

Quantification is also linked to control, particularly to the creation of nonhuman technologies that perform tasks in a given amount of time or make products of a given weight or size. Calculability is clearly linked to irrationality since, among other things, the emphasis on quantity tends to affect quality adversely.

Crucial to any discussion of calculability in contemporary society is the impact of the computer.[81] The proliferation of personal computers allows more of us to do more calculations with increasing speed. Many aspects of today's quantity-oriented society could not exist, or would need to be greatly modified, were it not for the modern computer. Although society undoubtedly was already moving toward increased calculability before computer technology advanced to its current level, computers have greatly expedited and extended that tendency.

Emphasizing Quantity Rather Than Quality of Products

McDonald's has always emphasized bigness; it and the other fast-food chains have a "bigger-is-better mentality."[82] For a long time, the most visible symbols of this emphasis were the large signs, usually beneath the even larger golden arches, touting the millions and, later, billions of hamburgers sold by McDonald's. This was a rather heavy-handed way of letting everyone know about McDonald's great success. (With the wide-scale recognition of their success in recent years, there is less need for McDonald's to be so obvious—hence the decline of such signs and the decrease in the size of the golden arches. Public protests against the size of the golden arches have also played a role in this.[83]) The mounting number of hamburgers sold not only indicated to potential customers that the chain was successful but also fostered the notion that the high quality of the burgers accounted for the immense sales. Hence, quantity appeared to equal quality.

The Fast-Food Industry:
Of "Big Bites" and "Super Big Gulps"

The emphasis on quantity is abundantly clear in the names of McDonald's offerings. The best-known example is the Big Mac. A large burger is considered desirable simply because consumers thus receive a large serving. Furthermore, consumers are led to believe that they are getting a great deal of food for a small price. Calculating consumers come away with the feeling that they are getting a good deal—and maybe even getting the best of McDonald's.

Many other fast-food restaurants mirror the emphasis by McDonald's on quantity. Burger King points out the quantity of meat in the "Whopper," the "Double Whopper," and even the "Triple Whopper." (Burger King's fish sandwich is "Big Fish.") Not to be outdone, Jack in the Box has its "Really Big

Chicken," Hardee's offers a "Monster Thickburger (2/3rds of a pound)," Kentucky Fried Chicken has a "Variety Big Box Meal" and "Double Down" (no bun, but two pieces of fried chicken, two pieces of bacon, and two pieces of cheese), and Taco Bell offers the "Big Bell Box Meal" with a half-pound of beans and rice.[84] Similarly, 7-Eleven proffers its customers a hot dog called the "Big Bite" and a large soft drink called the "Big Gulp," the even larger "Super Big Gulp," and the still-larger "Double Gulp."

Burger King also offers BK Stackers. The idea is that people can supersize their hamburger sandwich to include up to *three*(!) hamburger patties, two cheese slices, and three half slices of bacon. In one advertisement, the foreman at the "BK Stacker factory" is shown yelling "more meat" to workers struggling to produce a bigger burger. Of the product, Burger King says, "It's the flame-broiled meat lover's burger, and it's here to stay—no veggies allowed."[85] A meat lover who maximizes the Stacker will consume about 650 calories (the Triple Whopper has 1,140 calories), 1,020 milligrams of sodium (the Triple Whopper has a bit more), and about half of a day's allowance of saturated fat. Not to be outdone, Carl's Jr. has several varieties of its highly successful "Six Dollar Burger" (Hardee's has a "Six Dollar Thickburger"), although the "Double Six Dollar Burger," with about 75% of a day's suggested intake of calories (what does one do about eating for the rest of the day?), apparently went too far and was dropped.

Then there is Denny's well-known "Grand Slam Breakfasts." The emphasis is on not only the size of the meal but its low price. This is clear in one commercial in which a man proclaims, "I'm going to eat too much, but I'm never going to pay too much."[86] Of course, he and other consumers of Grand Slam Breakfasts are likely to "pay" in the long run (with their health) since it has many calories and large amounts of fat and sodium.

For years, McDonald's offered a "Super-Size" fries, 20% larger than a large order, and customers were urged to "supersize" their meals.[87] However, the uproar over Morgan Spurlock's documentary, *Supersize Me*, led McDonald's to drop the term, although it continues to offer menu items that emphasize large size. However, McDonald's continues to lag behind Hardee's and Carl's Jr., which are already offering half-pound burgers.[88]

Interestingly, the controversy over large sizes has led some fast-food restaurants to offer smaller portions of some menu items, although the emphasis is still on size rather than quality. Examples include Burger King's BK Mini Burgers and Jack in the Box's "Junior Bacon Cheeseburger." In this context, it is hard to know what to make of Hardee's "Little Thick Cheeseburger"—it is both little and big!

All this emphasis on quantity suggests that fast-food restaurants have little interest in communicating anything directly about quality.[89] Were they interested, they might give their products names such as "McTasty," "McDelicious," or "McPrime." But the fact is that typical McDonald's customers know they are not getting the highest-quality food: "No one . . . knows exactly what's in those hamburger patties. . . . Let's face it. Nobody thinks about what's between the bun at McDonald's. You buy, you eat, you toss the trash, and you're out of there like the Lone Ranger."[90]

Another observer has argued that people do not go to McDonald's for a delicious, pleasurable meal but, rather, to "refuel."[91] McDonald's is a place to fill our stomachs with lots of calories and carbohydrates so that we can move on to the next rationally organized activity. Eating to refuel is far more efficient than eating to enjoy a culinary experience.

The propensity for fast-food restaurants to minimize quality is well reflected in the sad history of Colonel Harland Sanders, the founder of Kentucky Fried Chicken. The quality of his cooking techniques and his secret seasoning (which his wife originally mixed, packed, and shipped herself) led to a string of about 400 franchised outlets by 1960. Sanders had a great commitment to quality, especially to his gravy: "To Sanders himself the supreme stuff of his art was his gravy, the blend of herbs and spices that time and patience had taught him. It was his ambition to make a gravy so good that people would simply eat the gravy and throw away 'the durned chicken.'"[92]

After Sanders sold his business in 1964, he became little more than the spokesman and symbol for Kentucky Fried Chicken. The new owners soon made clear their commitment to speed rather than quality: "The Colonel's gravy was fantastic, they agreed . . . but it was too complex, too time-consuming, too expensive. It had to be changed. It wasn't fast food." Ray Kroc, who befriended Colonel Sanders, recalls him saying, "That friggin' outfit . . . they prostituted every goddamn thing I had. I had the greatest gravy in the world and those sons of bitches they dragged it out and extended it and watered it down that I'm so goddamn mad."[93]

At best, what customers expect from a fast-food restaurant is modest but strong-tasting food—hence, the salty/sweet French fries, highly seasoned sauces, and saccharine shakes. Given such modest expectations of quality, customers do, however, have greater expectations of quantity. They expect to get a lot of food and to pay relatively little for it (or at least think we are getting a bargain).

Even the more upscale McDonaldized restaurant chains are noted for the size of their portions and the mediocrity of their food. Of the Olive Garden, one reviewer said, "But what brings customers in droves to this popular chain remains a mystery. The food defined mediocrity. Nothing was bad, but nothing was especially good, and it certainly isn't authentically Italian." Of course, the reason is quantity: "Portions . . . are large. . . . So you'll probably end up leaving stuffed, which is not to say satisfied."[94]

The Cheesecake Factory, begun in Beverly Hills in 1978 and now with nearly 150 restaurants, is another example of an upscale restaurant known for its huge portions (although many devotees consider its food to be higher in quality than, say, Olive Garden). Quantity is also in evidence in the menu, which contains over 200 highly varied items.[95] The cost of a meal at Cheesecake Factory is considered low for the quantity of food one consumes.

efficency?

Higher Education: Grades, Scores, Ratings, and Rankings

In education, most college courses run for a standard number of weeks and hours per week. Little attention is devoted to determining whether a given subject is best taught in a given number of weeks or hours per week. The focus seems to be on how many students ("products") can be herded through the system and what grades they earn rather than the quality of what they have learned and of the educational experience.

An entire high school or college experience can be summed up in a single number, the grade point average (GPA). Armed with their GPAs, students can take examinations with quantifiable results, such as the MCAT, LSAT, SAT, and GRE. Colleges, graduate schools, and professional schools can thus focus on three or four numbers in deciding whether or not to admit a student.

For their part, students may choose a university because of its rating. Is it one of the top 10 universities in the country? Is its physics department in the top 10? Are its sports teams usually top ranked? Most important, is it one of the top 10 party schools?

The increasing success of online universities is traceable to various quantitative factors. For one thing, courses are far less expensive than attending 4-year residential colleges, and credits can often be transferred to traditional colleges. For example, in one case, a student paid $750 for seven credits that would have cost $2,800 at a nearby college. In another, a master's degree in health care management from online Western Governor's University cost only $9,000 rather than the approximately $40,000 it would have cost at a traditional university.[96] At the online company, Straighterline, which has had more than a thousand students, "it costs $99 to register for a month, plus $39 a course—or a full 'freshman year' for $999."[97]

Potential employers may decide whether to hire graduates on the basis of their scores, their class ranking, and the ranking of the university from which they graduated. To increase their job prospects, students may seek to amass a number of different degrees and credentials with the hope that prospective employers will believe that, the longer the list of degrees, the higher the quality of the job candidate. Personal letters of reference, however important, are often replaced by standardized forms with quantified ratings (for example, "top 5% of the class," "ranks 5th in a class of 25").

The number of credentials a person possesses plays a role in situations other than obtaining a job. For example, people in various occupations can use long lists of initials after their names to convince prospective clients of their competence. (My BA, MBA, and PhD are supposed to persuade the reader that I am competent to write this book, although a degree in "hamburgerology" might be more relevant.) Said one insurance appraiser with ASA, FSVA, FAS, CRA, and CRE after his name, "The more [initials] you tend to put after your name, the more impressed they [potential clients] become."[98]

The sheer number of credentials, however, tells little about the competence of the person sporting them. Furthermore, this emphasis on quantity of credentials has led people to make creative use of letters after their names. For example, one camp director put "ABD" after his name to impress parents of prospective campers. As all academics know, this informal (and largely negative) label stands for "All But Dissertation" and is the label for people who have completed their graduate courses and exams but who have not written their dissertations. Also noteworthy here is the development of organizations whose sole reason for existence is to supply meaningless credentials, often through the mail.

The emphasis on quantifiable factors is common even among college professors (the "workers" if students are "products"). For example, at more and more colleges and universities, the students evaluate each course by answering questions with ratings of, for example, one to five. At the end of the semester, the professor receives what is in effect a report card with an overall teaching rating. Students have little or no room to offer qualitative evaluations of their teachers on such questionnaires. Although student ratings are desirable in a number of ways, they also have some unfortunate consequences. For example, they tend to favor professors who are performers, who have a sense of humor, or who do not demand too much from students. The serious professor who places great demands on students is not likely to do well in such rating systems, even though he or she may offer higher-quality teaching (for example, more profound ideas) than does the performer.

Quantitative factors are important not only in teaching but also in research and publication. The "publish or perish" pressure on academicians in many colleges and universities leads to greater attention to the quantity of their publications than to the quality. In hiring and promotion decisions, a resume with a long list of articles and books is generally preferred to one with a shorter list. Thus, an award-winning teacher was turned down for tenure at Rutgers University because, in the words of his department's tenure committee, his stack of publications was "not as thick as the usual packet for tenure."[99] The unfortunate consequence of this bias is publication of less than high-quality works, the rush to publication before a work is fully developed, or publication of the same idea or finding several times with only minor variations.

Another quantitative factor in academia is the ranking of the outlet in which a work is published. In the hard sciences, articles in professional journals receive high marks; books are less valued. In the humanities, books are of much higher value and sometimes more prestigious than journal articles. Being published by some publishers (for example, university presses) yields more prestige than being published by others (for example, commercial presses).

In sociology, for example, a formal ratings system assigns points for publication in professional journals. A publication in the prestigious *American Sociological Review* receives 10 points, the maximum in this system, but the far less prestigious (and, in order not to hurt anyone's feelings, fictional) *Antarctic Journal of Sociology,* with a readership primarily composed of penguins, receives only 1 point. By this system, the professor whose journal publications yield 340 points is supposed to be twice as "good" as one who earns only 170 points.

However, as is usually the case, such an emphasis on quantity tends not to relate to quality:

- It is highly unlikely that the quality of a professor's life's work can be reduced to a single number. In fact, it seems impossible to quantify the quality of an idea, theory, or research finding.
- This rating system deals only indirectly with quality. That is, the rating is based on the quality of the journal in which an article was published, not the quality of the article itself. No effort is made to evaluate the quality of the article or its contribution to the field. Poor articles can appear in the highest-ranking journals, excellent ones in low-ranking journals.
- The academician who writes only a few high-quality papers might not do well in this rating system. In contrast, someone who produces a lot of mediocre work could well receive a far higher score. This kind of system leads ambitious sociologists (and those in most other academic fields) to conclude that they cannot afford to spend years honing a single work because it will not pay off in their point score.

Any system that places so much emphasis on quantity of publications will lead to the production of a great deal of mediocre work.

The sciences have come up with another quantifiable measure in an effort to evaluate the quality of work: the number of times a person's work is cited in the work of other scholars. In many fields, Google Scholar is used, and it relies heavily on such citation counts.[100] In fact, Google now has a "citations gadget" to allow one to get an instantaneous citation count for virtually any scholar.[101] The assumption is that high-quality, important, and influential work is likely to be used and cited by other scholars. However, once again, the problem of evaluating quality arises. Can the influence of a person's academic work be reduced to a single number? Perhaps a few central uses of one scholar's ideas will influence the field more than many trivial citations of another scholar's work. Furthermore, the mere fact that a work is cited tells us nothing about how that work was used by other scholars. A worthless piece of work attacked by many people and thereby cited in their work would lead to many citations for its creator. Conversely, scholars may ignore a truly important piece of work that is ahead of its time, leading to a minuscule number of citations for the author. Google has tried to deal with this problem by, not surprisingly, coming up with another number—its h-index, which measures both number of publications and citations per publication.[102]

When he was president of Stanford University, Donald Kennedy announced a change in that university's policies for hiring, promoting, or granting tenure to faculty members. Disturbed by a report indicating "that nearly half of faculty members believe that their scholarly writings are merely counted—and not evaluated—when personnel decisions are made," Kennedy said, "I hope we can agree that the quantitative use of research output as a criterion for appointment or promotion is a bankrupt idea. . . . The overproduction of routine scholarship is one of the most egregious aspects of contemporary academic life: It tends to conceal really important work by sheer volume."[103]

To deal with this problem, Kennedy proposed to limit the extent to which the number of publications would be used in making personnel decisions. He hoped that the proposed limits would "reverse the appalling belief that counting and weighing are the important means of evaluating faculty research."[104] Despite protestations like this one, there is little evidence that there has been much progress in reducing the emphasis on quantity rather than quality in the academic world. In fact, in recent years, there has been a massive increase in the emphasis on quantitative factors in the British academic world. "League tables of universities are now produced grading research and teaching, as well as increased access by nonparticipating groups (ethnic minorities, working-class students). These make the system subject to quantitative, rather than the previous qualitative evaluations and therefore clearly calculable."[105]

Health Care: Patients as Dollar Signs

In profit-making medical organizations (the largest is HCA [Hospital Corporation of America]), physicians, along with all other employees, feel pressured to contribute to the corporation's profitability. Limiting the time with each patient and maximizing the number of patients seen in a day allow the corporation to reduce costs and increase profits. This emphasis on quantity can easily threaten the quality of medical care. Profits can also be increased by pushing doctors to turn down patients who probably can't pay the bills and to see only patients who have the kinds of diseases whose treatment is likely to yield large profits.

Following the lead of profit-making medical organizations, all medical bureaucracies are now pushing medicine in the direction of greater calculability. Even nonprofit medical organizations—for example, nonprofit hospitals and health maintenance organizations (HMOs, although most HMOs are for-profit, so perhaps a better example would be a local clinic)—increasingly employ professional managers and have sophisticated accounting systems.

The federal government, through Medicare, introduced the prospective payment and DRG (diagnostic-related groups)[106] programs, in which a set amount is reimbursed to hospitals for a given medical diagnosis, no matter how long the patient is hospitalized. Prior to 1983, the government paid whatever "reasonable" amount it was billed. Outside agencies have also grown increasingly concerned about spiraling medical costs and have sought to deal with the problem by limiting what they will pay for and how much they will pay for it. A third-party payer (an insurer) might thus refuse to pay for certain procedures or for hospitalization or perhaps pay only a given amount.

Doctors, who have traditionally placed quality of patient care above all else (at least ideally), often complain about the new emphasis on calculability. At least one physicians union engaged in a strike centered on issues such as required number of visits, number of patients seen, and an incentive system tying physician salaries to productivity. As one physician union leader put it, however romantically, doctors are "the only ones who think of patients as individuals . . . not as dollar signs."[107]

Sports: Nadia Comaneci Scored Exactly 79.275 Points

The quality of various sports has been altered by, perhaps even sacrificed to, calculability. For instance, the nature of sporting events has been changed by the needs of television.[108] Because teams in many sports earn a large part of their revenue from television contracts, they will sacrifice the interests of paid spectators, even compromise the games themselves, to increase their television income.

A good example is the so-called TV timeout. In the old days, commercials occurred during natural breaks in a game—for example, during a timeout called by one of the teams, at halftime, or between innings. But these breaks were too intermittent and infrequent to bring in the increasingly large fees advertisers were willing to pay. Now regular TV timeouts are scheduled in sports such as football and basketball. The owners of sports franchises may be maximizing their incomes from advertising, but the momentum of a team may be lost because of an inopportune TV timeout. These timeouts not only alter the nature of some sports but may even affect the outcome of games. Also, these timeouts interrupt the flow of the game for the fans who watch in person (and pay high ticket prices for the privilege). The fans at home can at least watch the commercials; the spectators at the games have little to watch until the commercial ends and the game resumes. But the owners consider such negative effects on the quality of the game insignificant compared with the economic gain from increased advertising.

Clearly, though, sports continue to place a premium on the quality of both individual and team performance: for instance, the offensive skills of basketball stars Kobe Bryant and LeBron James and the teamwork of the Pittsburgh Steelers, who won the 2009 Super Bowl. At the same time, quantitative factors have always been enormously important in sports. In many cases, quality is directly related to quantity: the better the performance, the higher the score, and the greater the number of victories. Over the years, however, the emphasis on the quantifiable attributes of sports has increased:

> Modern sports are characterized by the almost inevitable tendency to transform every athletic feat into one that can be quantified and measured. The accumulation of statistics on every conceivable aspect of the game is a hallmark of football, baseball, hockey, and of track and field too, where the accuracy of quantification has, thanks to an increasingly precise technology, reached a degree that makes the stopwatch seem positively primitive.[109]

Even a highly aesthetic sport such as gymnastics has been quantified: "How can one rationalize and quantify a competition in gymnastics, in aesthetics? . . . Set up an interval scale and a panel of judges and then take the arithmetic mean of the subjective evaluations. . . . Nadia Comaneci [an Olympic champion in gymnastics] scored exactly 79.275 points . . . neither more nor less."[110]

The growing emphasis on quantity can sometimes adversely affect the quality of play in a sport. For example, the basketball star motivated by a need to stand out individually and score as many points as possible may negatively affect the play of individual teammates and the team's overall performance. The quality of play is even more compromised by owners' attempts to maximize points scored. Basketball, for example, was once a rather leisurely game in which a team could take as long as necessary to bring the ball down the court and get a player into position to take a good shot. Basketball fans enjoyed the strategies and maneuvers employed by the players. Toward the end of the game, a team holding a slim lead could attempt to "freeze" the ball—that is, not risk missing a shot and thereby giving their opponents a chance to take possession of the ball.

Several decades ago, however, the leadership of collegiate and professional basketball decided that fans raised in the McDonald's era wanted to see faster games and many more points scored. In other words, fans wanted from basketball what they got from their fast-food restaurant: great speed and large quantities. Hence, in college games, offensive teams were limited to 35 seconds in which to attempt a shot; in professional games, 24-second time clocks were established. Although the "run-and-shoot" style of play generated by the time clocks has created faster-paced, higher-scoring games, it may have adversely affected the quality of play. No longer is there much time for the maneuvers and strategies that made the game so interesting to "purists." But a run-and-shoot style of basketball fits in well with the McDonaldized "eat-and-move" world of dinners purchased at drive-through windows and consumed on the run. In recent years, however, professional basketball teams have developed interesting new defensive strategies, and the scores of games have plummeted. There is now interest in making further changes in the game to reduce the advantages of the defensive team and, thereby, once again inflate the scores.

Similarly, baseball owners decided long ago that fans prefer to see high-scoring games with lots of hits, home runs, and runs scored rather than the kind of game favored by "purists": pitchers' duels in which the final score might be 1–0. They took a number of steps to increase the number of runs scored. Livelier baseballs travel farther than old-fashioned "dead balls." In some baseball parks (the new Yankee Stadium, opened in 2009, is notable in this regard), outfield fences have been brought closer to home plate to increase the number of home runs.

The designated hitter, found in the American League but not in the more traditional National League, is the most important effort to increase hits and runs. Instead of the often weak-hitting pitcher taking his turn at bat, someone whose main (and sometimes only) skill is batting replaces him. Designated hitters get more hits, hit more home runs, and help produce more runs than pitchers who are allowed to bat.

Although use of the designated hitter in American League baseball has undoubtedly increased the number of runs scored, it has also affected, perhaps adversely, the quality of the game (which is why the National League has steadfastly refused to adopt the designated hitter). For example, when pitchers bat in certain situations, they often employ a sacrifice bunt, a very artful practice intended to advance a runner already on base. But a designated hitter rarely sacrifices an at-bat by bunting. In addition, pinch hitters play less of a role when designated hitters replace weak-hitting pitchers.[111] Finally, if there is less need to pinch hit for them, starting pitchers can remain in games longer, which reduces the need for relief pitchers.[112] In these and other ways, baseball is a different game when a designated hitter is employed. In other words, the quality of the game has changed, some would say for the worse, because of the emphasis on quantity.

The term *extreme* has been associated with all sorts of things (food [e.g., 7-Eleven's "X-treme Gulp," Denny's "Extreme Grand Slam Breakfast"], candy, breath mints), although most typically with an array of sports activities. These include *extreme* snowboarding, skateboarding, BMX riding, and even bowling.[113] The use of this term implies that these sports offer something "more" than their traditional versions. Usually, the "more" translates into that which is faster, riskier, more innovative, more creative, and more outrageous. Most generally, extreme sports are associated with that which is more nonconformist, more dare-devilish, and, in a word, "cooler" than traditional sports.

Politics: There Were No Sound Bites
in the Lincoln-Douglas Debate

The political sector offers a number of revealing examples of the emphasis on calculability—for example, the increasing importance of polls in political campaigns.[114] Candidates and incumbents, obsessed by their ratings in political polls, often adjust their positions on issues or the actions they take on the basis of what pollsters say will increase (or at least not lower) their rankings. How a specific political position affects ratings can become more important than whether the politician genuinely believes in it.

Television has affected politics in various ways. It has led to shorter conventions, for one thing, as well as shorter political speeches. In the famous Lincoln-Douglas debates of 1858, the candidates "spoke for ninety minutes each on a single topic: the future of slavery in the territories."[115] Prior to television, political speeches on radio at first often lasted an hour; by the 1940s, the norm had dropped to 30 minutes. In the early years of television, speeches also lasted about half an hour, but because political campaign speeches became more tailored to television coverage and less to the immediate audience, speeches have grown shorter, less than 20 minutes on the average. By the 1970s, speeches had been largely replaced by the 60-second advertisement. In today's televised presidential debates, candidates have a minute or two to offer their position on a given issue.

News reports of political speeches have shrunk to fit the visual demands of television as well. By the 1984 presidential campaign, only about 15 seconds of a speech would likely find its way onto a national news program. Four years later, speaking time on such reports shrank to only 9 seconds.[116] As a result, political speechwriters concentrate on creating 5- or 10-second "sound bites" that are likely to be picked up by the national networks. This emphasis on length has clearly reduced the quality of public political speeches and, therefore, the quality of public discourse on important political issues. (However, YouTube is countering this trend. Barack Obama won the presidency in 2008 partly because of his savvy use of such Web 2.0 technologies. YouTube allowed him to stream more content in much more depth on the Web, with millions of people viewing, for example, his famous speech on race online;[117] see Chapter 7 for a discussion of the ways in which Web 2.0 tends to lean toward deMcDonaldization.)

Reducing Production and Service to Numbers

The emphasis on the number of sales made and the size of the products offered are not the only manifestations of calculability in fast-food restaurants.

The Fast-Food Industry: Hustle, and a Precooked Hamburger Measures Exactly 3.875 Inches

Another example is the great emphasis on the speed with which a meal can be served. In fact, Ray Kroc's first outlet was named McDonald's Speedee Service Drive-In. At one time, McDonald's sought to serve a hamburger, shake, and French fries in 50 seconds. The restaurant made a great breakthrough in 1959 when it served a record 36 hamburgers in 110 seconds.

Many other fast-food restaurants have adopted McDonald's zeal for speed. Burger King, for example, seeks to serve a customer within 3 minutes of entering the restaurant.[118] The drive-through window drastically reduces the time required to process a customer through a fast-food restaurant. Speed is obviously a quantifiable factor of monumental importance in a fast-food restaurant.

Speed is even more important to the pizza delivery business. At Domino's, the mantra is "Hustle! Do It! Hustle! Do It! Hustle! Do It!" and the "Domino's goal is eight minutes out the door."[119] Not only does the number sold depend on how quickly the pizzas can be delivered, but also a fresh pizza must be transported quickly to arrive hot. Special insulated containers now help keep the pizzas hot longer. This emphasis on rapid delivery has caused several scandals, however; pressure to make fast deliveries has led young delivery people to become involved in serious and sometimes fatal automobile accidents.

Still another aspect of the emphasis on quantity lies in the precision with which every element in the production of fast food is measured. At frozen yogurt franchises, containers are often weighed to be sure they include the correct quantity of frozen yogurt, whereas in old-fashioned ice cream parlors, the attendants simply filled a container to the brim. McDonald's itself takes great care in being sure that each raw McDonald's hamburger weighs exactly 1.6 ounces—10 hamburgers to a pound of meat. The precooked hamburger measures precisely 3.875 inches in diameter and the bun exactly 3.5 inches. McDonald's invented the "fatilyzer" to ensure that its regular hamburger meat had no more than 19% fat.[120] Greater fat content would lead to greater shrinkage during cooking and prevent the hamburger from appearing too large for the bun. Grilling eight hamburger patties at a time takes 38 seconds.[121] The French fry scoop helps make sure that each package has about the same number of fries. The automatic drink dispensers ensure that each cup gets the correct amount of soft drink, with nothing lost to spillage. The average cash register transaction takes 12 seconds.[122]

Arby's has reduced the cooking and serving of roast beef to a series of exact measurements.[123] All roasts weigh 10 pounds at the start. They are roasted at 200°F for 3.5 hours until the internal temperature is 135°F. They are then allowed to cook in their own heat for 20 minutes more until the internal temperature is 140°F. By following these steps and making these measurements, Arby's doesn't need a skilled chef; virtually anyone who can read and count can cook an Arby's roast beef. When the roasts are done, each weighs between 9 pounds, 4 ounces and 9 pounds, 7 ounces. Every roast beef sandwich contains 3 ounces of meat, allowing Arby's to get 47 sandwiches (give or take one) from each roast.

Burger King has also quantified quality control. Hamburgers must be served within 10 minutes of being cooked. French fries may stand under the heat lamp for no more than 7 minutes. A manager is allowed to throw away 0.3% of all food.[124]

The performance of fast-food restaurants is also assessed quantitatively, not qualitatively. At McDonald's, for example, central management judges the performance of each restaurant "by 'the numbers': by sales per crew person, profits, turnover, and QSC [Quality, Service, Cleanliness] ratings."[125]

While the fast-food restaurant has greatly increased the emphasis on calculability, it had many precursors, including the original *Boston Cooking School Cook Book* (1896), in which Fannie Farmer emphasized precise measurement and, in the process, helped rationalize home cooking: "She . . . changed American kitchen terminology from 'a pinch' and 'a dash' and 'a heaping spoonful' . . . to her own precise, standardized, scientific terms, presenting a model of cooking that was easy, reliable, and could be followed even by inexperienced cooks."[126] For those who overeat on food based on Fannie Farmer recipes, or more likely these days on fast food, there are the exercise chains built on a similar emphasis on quantification. The success of Curves is based on offering a simple set of machine-based exercises that can be completed in just 30 minutes.

The Workplace: A Penny the Size of a Cartwheel

With scientific management, Taylor intended to transform everything work related into quantifiable dimensions. Instead of relying on the worker's "rule of thumb," scientific management sought to develop precise measurements of how much work was to be done by each and every motion of the worker. Everything that could be reduced to numbers was then analyzed using mathematical formulas.

Calculability was clearly an aim when Taylor sought to increase the amount of pig iron a worker could load in a day: "We found that this gang were loading on the average about 12½ long tons per man per day. We were surprised to find, after studying the matter, that a first-class pig iron handler ought to handle between 47 and 48 long tons per day, instead of 12½ tons."[127] To try to nearly quadruple the workload, Taylor studied the way the most productive workers, the "first-class men," operated. He divided their work into its basic elements and timed each step with a stopwatch, down to hundredths of a minute.

On the basis of this careful study, Taylor and his associates developed the one best way to carry pig iron. They then found a worker they could motivate to work this way—Schmidt, who was able and ambitious and to whom, as one coworker said, a penny looked to be "about the size of a cart-wheel." Schmidt indicated that he wanted to be a "high-priced man." Taylor used a precise economic incentive: $1.85 per day, rather than the usual $1.15, if Schmidt agreed to work exactly the way Taylor told him to. After careful training and supervision, Schmidt successfully worked at the faster pace (and earned the higher pay); Taylor then selected and trained other employees to work the same way.

Schmidt and his successors, of course, were being asked to do about 3.6 times the normal amount of work for an approximately 60% increase in pay. For Taylor, "the pig iron handler with his 60-percent increase in wages is not an object for pity but rather a subject for congratulations."[128]

Also illustrative of the impact of calculability on industry is the famous case of the Ford Pinto.[129] Because of competition from the manufacturers of small foreign cars, Ford rushed the Pinto into production, even though preproduction tests had indicated that its fuel system would rupture easily in a rear-end collision. The expensive assembly-line machinery for the Pinto was already in place, so Ford decided to go ahead with the production of the car without any changes. Ford based its decision on a quantitative comparison. The company estimated that the defects would lead to 180 deaths and about the same number of injuries. Placing a value, or rather a cost, on them of $200,000 per person, Ford decided that the total cost from these deaths and injuries would be less than the $11 per car it would cost to repair the defect. Although this calculation may have made sense from the point of view of profits, it was an unreasonable decision in that human lives were sacrificed and people were maimed in the name of lower costs and higher profits. This example is only one of the more extreme of a number of such decisions made daily in a society undergoing McDonaldization.

4

Predictability and Control

T he focus in this chapter is on the two other dimensions of McDonaldization: predictability and control.

Predictability: It Never Rains on Those Little Houses on the Hillside

In a rationalized society, people prefer to know what to expect in most settings and at most times. They neither desire nor expect surprises. They want to know that when they order their Big Mac today, it will be identical to the one they ate yesterday and the one they will eat tomorrow. People would be upset if the special sauce was used one day but not the next or if it tasted differently from one day to the next. They want to know that the McDonald's franchise they visit in Des Moines, Los Angeles, Paris, or Beijing will appear and operate much the same as their local McDonald's. To achieve predictability, a rationalized society emphasizes discipline, order, systematization, formalization, routine, consistency, and methodical operation.

From the consumer's point of view, predictability makes for much peace of mind in day-to-day dealings. For workers, predictability makes tasks easier. In fact, some workers prefer effortless, mindless, repetitive work because, if nothing else, it allows them to think of other things, even daydream, while they are doing their tasks.[1] For managers and owners, too, predictability makes life easier: It helps them manage both workers and customers and aids in anticipating needs for supplies and materials, personnel requirements, income, and profits.

Predictability, however, has a downside. It has a tendency to turn everything—consumption, work, management—into mind-numbing routine.

Creating Predictable Settings

Motel chains, like fast-food restaurants, are pioneers in the rationalization process. The most notable are Best Western, founded in 1946 (more than 4,000 hotels in 90 countries),[2] and Holiday Inn (and Holiday Inn Express), which started in 1952. Holiday Inn is now part of the Inter-Continental Hotels Group, which claims to be the world's largest hotel chain with about 4,400 hotels in more than 100 countries; about 3,300 of them carry the Holiday Inn logo.[3] By the late 1950s, about 500 Howard Johnson's restaurants (only 3 remain) were scattered around the United States, many with standardized motels attached to them. Unlike other motel chains, Howard Johnson's has stagnated, but it still has 475 motels in 13 countries; it is now part of Wyndham Worldwide.[4] These three motel chains, and others, opened in anticipation of the massive expansion of highways and highway travel. Their ability to bring consistency to the motel and hotel industry was the basis of their success and has been widely emulated.

Motel Chains: "Magic Fingers" but No Norman Bates

Before the development of such chains, motels were very diverse and highly unpredictable. Run by local owners, every motel was different from every other. Because the owners and employees varied from one locale to another, guests could not always feel fully safe and sleep soundly. One motel might be quite comfortable, even luxurious, but another might well be a hovel. People could never be sure which amenities would be present—soap, shampoo, telephones, radio (and later television), air-conditioning, and please don't forget the much-loved "Magic Fingers" massage system. Checking into a motel was an adventure; a traveler never knew what to expect.

In his classic thriller, *Psycho* (1960), Alfred Hitchcock beautifully exploited the anxieties about old-fashioned, unpredictable motels. The motel in the movie was creepy but not as creepy as its owner, Norman Bates. Although it offered few amenities, the Bates Motel room did come equipped with a peephole (something most travelers could do without) so that Norman could spy on his victims. Of course, the Bates Motel offered the ultimate in unpredictability: a homicidal maniac and a horrible death to unsuspecting guests.

Although very few independently owned motels actually housed crazed killers, all sorts of unpredictabilities confronted travelers at that time. In contrast, the motel chains took pains to make their guests' experience predictable. They developed tight hiring practices to keep "unpredictable" people from managing or working in them. Travelers could anticipate that a motel equipped with the then-familiar orange and green Holiday Inn sign (now gone the way of McDonald's oversized golden arches) would have most, if not all, the amenities they could reasonably expect in a moderately priced motel. Faced with the choice between a local, no-name motel and a Holiday Inn, many travelers preferred the predictable—even if it had liabilities (the absence of a personal touch, for example). The success of the early motel chains has led to many imitators, such as Ramada Inn and Rodeway Inn (now part of Choice Hotels International).

The more price-conscious chains—Super 8, Days Inn, and Motel 6—are, if anything, even more predictable.[5] The budget motel chains are predictably barren; guests find only the minimal requirements. But they expect the minimum, and that's what they get. They also expect, and receive, bargain-basement prices for the rooms.

The Fast-Food Industry:
Thank God for Those Golden Arches

The fast-food industry quickly adopted and perfected practices pioneered by, among other precursors, the motel chains. In fact, Robin Leidner argues that "the heart of McDonald's success is its uniformity and predictability, . . . [its] relentless standardization." She argues that "there is a McDonald's way to handle virtually every detail of the business, and that doing things differently means doing things wrong."[6] Although McDonald's allows its franchisees and managers to innovate, "The object is to look for new innovative ways to create an experience that is exactly the same no matter what McDonald's you walk into, no matter where it is in the world."[7]

Like the motel chains, McDonald's (and many other franchises) devised a large and garish sign that soon became familiar to customers. McDonald's "golden arches" evoke a sense of predictability: "Replicated color and symbol, mile after mile, city after city, act as a tacit promise of *predictability* and stability between McDonald's and its millions of customers, year after year, meal after meal" (italics added).[8] Even though the signs are now smaller and less obtrusive, they still conjure a feeling of familiarity and predictability for consumers. Furthermore, each McDonald's presents a series of predictable elements—counter, menu marquee above it, "kitchen" visible in the background, tables and uncomfortable seats, prominent trash bins, drive-through windows, and so on.

This predictable setting appears not only throughout the United States but also in many other parts of the world. Thus, homesick American tourists can take comfort in the knowledge that nearly anywhere they go, they will likely run into those familiar golden arches and the restaurant to which they have become so accustomed. Interestingly, even many non-Americans now take comfort in the appearance of a familiar McDonald's restaurant when they journey to other countries, including the United States itself.

This kind of predictability is much the same in all fast-food chains. Starbucks has its familiar green and white sign, mermaid (recently stream-lined), counter for placing and paying for orders, display case filled mainly with pastries, and a separate counter where specialty drinks are made and retrieved. Then there is Kentucky Fried Chicken with its cartoonish image of Colonel Sanders; Wendy's little girl with red pigtails; Taco Bell with its logo that features a bell; Papa John's red, white, and green logo designed to remind us of Italy by replicating the colors of the Italian flag; and In-N-Out Burger's sweeping yellow arrow.

Other Settings: E.T. Can't Find His Home

As work settings, bureaucracies are far more predictable than other kinds of organizations. They engender predictability in at least three important ways. First, employees occupy "offices" and are expected to live up to expec-tations associated with them. Second, a bureaucracy has a clear hierarchy of offices so that people know from whom to take orders and to whom they may give orders. Third, virtually everything in a bureaucracy exists in written form. Thus, those who read the organization's rules and regulations know what can be expected. Handling an issue often involves little more than filling out a particular form. These days, such forms are on the computer as are all mes-sages in the form of e-mails and their attachments, text messages, tweets, and so on.

Megachurches are defined as churches with more than 2,000 members (the largest one is Lakewood Church in Houston, Texas, with an average attendance of 43,500).[9] The 1,380 megachurches in the United States[10] play a central role in the McDonaldization of religion. For example, a McLean, Virginia, megachurch is planning to expand by opening nine satellite loca-tions around the Washington beltway (a number of megachurches now have satellite locations). While there might be some variation, some customization from one satellite to another (live music, local pastors), all hear a portion of the same sermon at the same time via television. This serves to bring a pol-ished, predictable, and "branded" performance to every location served by the megachurch.[11]

Modern suburban housing also demonstrates the predictability of settings in a McDonaldized society. A famous folk song, in fact, characterizes the sub-urbs as

Little boxes on the hillside,

Little boxes made of ticky-tacky,

Little boxes, little boxes,

Little boxes all the same.[12]

In many suburban communities, interiors and exteriors are little different from one house to another. Although some diversity exists in the more expensive developments, many suburbanites could easily wander into someone else's house and not realize immediately that they were not in their own home.

Furthermore, the communities themselves look very much alike. Rows of saplings held up by posts and wire are planted to replace mature trees ripped out of the ground to allow for the more efficient building of houses. Similarly, hills are often bulldozed to flatten the terrain. Streets are laid out in familiar patterns. With such predictability, suburbanites may well enter the wrong suburban community or get lost in their own community.

Several of Steven Spielberg's early movies take place in these rationalized suburbs. Spielberg's strategy is to lure viewers into this highly predictable world and then hit them with a highly unpredictable event. For example, in *E.T.* (1982), an extraterrestrial wanders into a suburban development of tract houses and is discovered by a child there who, until that point, has lived a highly predictable suburban existence. The unpredictable E.T. eventually disrupts not only the lives of the child and his family but also the entire community. Similarly, *Poltergeist* (1982) takes place in a suburban household, with evil spirits disrupting its predictable tranquility. The great success of several of Spielberg's movies may be traceable to people's longing for some unpredictability, even if it is frightening and menacing, in their increasingly predictable suburban lives.

The Truman Show (1998) takes place in a community completely controlled by the director of a television show. The movie can be seen as a spoof of, and attack on, the Disneyesque "planned communities" springing up throughout the United States. These are often more upscale than the typical suburban community. The leading example of a planned community is, not surprisingly, Disney's town of Celebration, Florida. Potential homeowners must choose among approved options and are strictly limited in what they can do with their homes and property.[13] These communities go farther than traditional suburban developments in seeking to remove all unpredictability from people's lives.

Another 1998 movie, *Pleasantville,* depicts a 1950ish community that is tightly controlled and characterized by a high degree of conformity and uniformity. This is reflected in the fact that everything is depicted in black and white. As the story unfolds, however, things grow less and less predictable, and color is gradually introduced. In the end, a far more unpredictable Pleasantville is shown in full color. The remake of *Fun With Dick and Jane* (2005) as well as *Revolutionary Road* (2008) deal with couples who have, and then gradually lose, the seemingly ideal suburban home and lifestyle.

Although they strive hard to be as predictable as possible, some of the more recent chains are finding a high degree of predictability an elusive goal. For example, haircutting franchises such as Hair Cuttery (750 salons in 16 states[14]) cannot offer a uniform haircut because every head is slightly different and every barber or hairdresser operates in a slightly idiosyncratic fashion. To reassure the anxious customer longing for predictability, MasterCuts, Great Clips, Hair Cuttery, and other haircutting franchises offer a common logo and signs, similar shop setup, and perhaps a few familiar products.

Scripting Interaction With Customers

Even more reminiscent of *The Truman Show* are the scripts that McDonaldized organizations prepare for employees. In the movie, every character's interaction with Truman followed a script provided by the director to make Truman's actions more predictable.

The Fast-Food Industry: "Howdy, Pardner" and "Happy Trails"

Perhaps the best-known scripts at McDonald's are "Do you want fries with that?" and "Would you like to add an apple pie to that order?"[15] Such scripts help create highly predictable interactions between workers and customers. While customers do not follow scripts, they tend to develop their own simple recipe-like responses to the employees of McDonaldized systems.

The Roy Rogers chain used to have its employees, dressed in cowboy and cowgirl uniforms, say "Howdy, pardner" to every customer about to order food. After paying for the food, people were sent on their way with "Happy trails." The repetition of these familiar salutations visit after visit was a source of great satisfaction to regulars at Roy Rogers. Many people (including me) felt a deep personal loss when Roy Rogers ceased this practice. However, in a McDonaldized society, other types of pseudo interactions are increasingly the norm. Consumers have come to expect and maybe even like them, and they might even look back on them longingly when interactions with their favorite robot are all they can expect on their visits to fast-food restaurants.

Like all other aspects of McDonaldization, scripts can have positive functions. For example, scripts can be a source of power to employees, enabling them to control interaction with customers. Employees can fend off unwanted or extraordinary demands merely by refusing to deviate from the script.

Employees can also use their routines and scripts to protect themselves from the insults and indignities frequently heaped on them by the public. Employees can adopt the view that the public's hostility is aimed not at them personally but at the scripts and those who created them. Overall, rather than being hostile toward scripts and routines, McDonald's workers often find them useful and even satisfying.[16]

However, employees and customers sometimes resist scripts (and other routines). As a result, what those who give and receive services actually say "is never entirely predictable."[17] People do not yet live in an iron cage of McDonaldization. In fact, they are unlikely ever to live in a totally predictable, completely McDonaldized world.

Nonetheless, what McDonald's workers could do to exert some independence in their work is hardly overwhelming. For example, they could say all sorts of things that are not in the script, go a "bit" beyond the routine by providing extra services or exchanging pleasantries, and withhold smiles, act a bit impatient or irritated, or refuse to encourage the customer to return. These all seem like very small deviations from an otherwise highly routinized workday.

As in the case of workers, customers also gain from scripts and routines: "Routinization can provide service-recipients with more reliable, less expensive, or speedier service, can protect them from incompetence, can minimize the interactive demands on them, and can clarify what their rights are." Such routines help guarantee equal treatment of all customers. Finally, routinization can help "establish a floor of civility and competence for which many customers have reason to be grateful."[18] Some customers appreciate the polite, ritualized greetings they encounter at McDonaldized businesses.

However, there are exceptions. Some customers may react negatively to employees mindlessly following scripts, who seem "unresponsive" or "robot-like."[19] Arguments can ensue, and angry customers may even leave without being served. In a classic scene from *Five Easy Pieces* (1970), Jack Nicholson's character stops at a diner and encounters a traditional greasy-spoon waitress, who is following a script. He cannot get an order of toast, although he could order a sandwich made with toast. Nicholson's character reacts even more strongly and negatively to the unresponsive script than he does to the surly waitress.

The fake friendliness of scripted interaction reflects the insincere camaraderie ("have a nice day") that characterizes not only fast-food restaurants but also all other elements of McDonaldized society, a camaraderie used to lure customers and keep them coming back. For example, before his death, TV screens were saturated with scenes of Wendy's owner and founder, Dave Thomas, extending a "personal invitation" to join him for a burger at his restaurant.[20]

Other Settings: Even the Jokes Are Scripted

The fast-food industry is far from the only place where we are likely to encounter scripted interaction. Telemarketing is another setting that usually provides scripts that workers must follow unerringly. The scripts are designed to handle most foreseeable contingencies. Supervisors often listen in on solicitations to make sure employees follow the correct procedures. Employees who fail to follow the scripts, or who do not meet the quotas for the number of calls made and sales completed in a given time, may be fired summarily.

Robin Leidner details how Combined Insurance tried to make life insurance sales predictable: "The most striking thing about Combined Insurance's training for its life insurance agents was the amazing degree of standardization for which the company was striving. The agents were told, in almost hilarious detail, what to say and do." In fact, a large portion of the agents' sales pitch was "supposed to be memorized and recited as precisely as possible." One of the trainers told of a foreign salesman whose English was poor, "He learned the script phonetically and didn't even know what the words meant. . . . He sold twenty applications on his first day and is now a top executive."[21] The insurance agents were even taught the company's standard joke, as well as "the Combined shuffle"—standardized movements, body carriage, and intonation.

McDonald's relies on external constraints on its workers, but Combined Insurance attempts to transform its workers. Combined workers are supposed to embrace a new self (a "McIdentity");[22] in contrast, McDonald's employees are expected to suppress their selves. This dissimilarity is traceable to differences in the nature of the work in the two settings. Because McDonald's workers perform their tasks within the work setting, they can be controlled by external constraints. In contrast, Combined Insurance salespeople work door-to-door, with most of their work done inside customers' homes. Because external constraints do not work, Combined tries to change the agents into the kind of people the company wants. Despite the efforts to control their personalities, however, the insurance agents retain some discretion as well as a sense of autonomy. Although control over Combined workers goes deeper, McDonald's workers are still more controlled because virtually all decision making is removed from their jobs. Leidner concludes, "No detail is too trivial, no relationship too personal, no experience too individual, no manipulation too cynical for some organization or person, in a spirit of helpfulness or efficiency, to try to provide a standard, replicable routine for it."[23]

Making Employee Behavior Predictable

The assembly line enhanced the likelihood of predictable work and products. The problem with the alternative is that the steps a craftsperson takes are somewhat unpredictable, varying from person to person and over time. Small but significant differences in the finished products crop up, and they lead to unpredictabilities in the functioning and quality of the goods. For instance, one car produced by a craftsperson would run far better or be much less prone to breakdowns than a car produced by another; cars produced on an assembly line are far more uniform. Realizing the benefits to be gained from predictable worker performance, many nonmanufacturing industries now have highly developed systems for making employee behavior more routine.

The Fast-Food Industry: Even Hamburger University's Professors Behave Predictably

Because interaction between customer and counter person in the fast-food restaurant is limited in length and scope, it can be largely routinized. McDonald's thus has a series of regulations that employees must follow in their dealings with customers. There are, for example, seven steps to window service: greet the customer, take the order, assemble the order, present the order, receive payment, thank the customer, and ask for repeat business.[24] Fast-food restaurants also seek to make other work as predictable as possible. For example, all employees are expected to cook hamburgers in the same, one-best way. In other words, "Frederick Taylor's principles can be applied to assembling hamburgers as easily as to other kinds of tasks."[25]

Employees follow well-defined steps in preparing various foods. The following are those involved in preparing French fries in much of the fast-food industry: Open the bag of (frozen) fries; fill the basket half full; put the basket in the deep fry machine; push the timer; remove the basket from the fryer and dump the fries into holding trays; shake salt on the fries; push another timer; discard the fries that are unsold after several minutes; look at the computer screen to see what size fries the next customer requires; fill another fry container and put it in a holding bin in preparation for preparing the next batch of fries.

Fast-food restaurants try in many ways to make workers look, act, and think more predictably.[26] All employees must wear uniforms and follow dress codes for things such as makeup, hair length, and jewelry. Training programs are designed to indoctrinate the worker into a "corporate culture,"[27] such as the McDonald's attitude and way of doing things. Highly detailed manuals spell out, among other things, "how often the bathroom must be cleaned to the temperature of grease used to fry potatoes . . . and what color nail polish to wear."[28] Finally, incentives (awards, for example) are used to reward employees who behave properly and disincentives, ultimately firing, to deal with those who do not.

To help ensure predictable thinking and behavior among restaurant managers, McDonald's has them attend its central Hamburger University in Oak

Brook, Illinois, or one of its branches throughout the United States and the world (Sydney, Munich, London, Tokyo, Hong Kong, and Brazil).[29] Hamburger University was begun in 1961, has about 5,000 "students" a year, and has graduated more than 80,000 of them. It offers separate "tracks" for crew, restaurant managers, mid-level managers, and executives. Even the 19 full-time "professors" at Hamburger University behave predictably because "they work from scripts prepared by the curriculum development department."[30] Trained by such teachers, managers internalize McDonald's ethos and its way of doing things. As a result, in demeanor and behavior, McDonald's managers are hard to distinguish from one another. More important, because the managers train and oversee workers to help make them behave more predictably, managers use elaborate corporate guidelines that detail how virtually everything is to be done in all restaurants. McDonald's central headquarters periodically sends forth "undercover" inspectors to be sure that these guidelines are being enforced. These inspectors also check to see that the food meets quality control guidelines.

Other Settings: That Disney Look

Amusement parks have adopted many similar techniques. Disney has developed detailed guidelines about what its employees should look like (the "Disney look") and how they should act. Disney has assembled a long list of "dos" and "don'ts" for employees. Female "cast members" (a Disney euphemism for its park employees) may not use beads as part of a braided hairstyle, shave their head or eyebrows, or use extreme hair coloring. Only neutral nail polish is allowed, and fingernails must not be more than a quarter of an inch in length. Neat mustaches are permitted for male hosts, but beards and goatees are not. Male cast members are allowed to shave their heads but not their eyebrows. The list goes on and on.[31]

Disney is not alone among amusement parks in the effort to make employee behavior predictable. At Busch Gardens, Virginia, "a certain amount of energy is devoted to making sure that smiles are kept in place. There are rules about short hair (for the boys) and no eating, drinking, smoking, or straw chewing on duty (for everyone)."[32] Not only do the employees at Busch Gardens all look alike, but they are also supposed to act alike: "Controlled environments hinge on the maintenance of the right kind of attitude among the lower echelons. 'It is kind of a rah-rah thing. We emphasize cleanliness, being helpful, being polite'. . . . Consequently, there is a lot of talk at Busch Gardens about . . . keeping people up and motivated . . . there are contests to determine who has the most enthusiasm and best attitude" (italics added).[33] Such techniques ensure that visitors to Busch Gardens and parks like it can expect to see and deal with highly predictable employees throughout their visit.

Creating Predictable Products and Processes

The drive for increased predictability extends, not surprisingly, to the goods and services being sold and the methods used to produce and deliver them. Consider the uniformity that characterizes the chain stores (Apple, Banana Republic, Foot Locker, Old Navy, United Colors of Benetton, Victoria's Secret, etc.) dominating virtually all malls. Few of the products are unique—indeed, many are globally available brand names—and procedures for displaying merchandise, greeting customers, ringing up purchases, and so on are amazingly similar.

The Fast-Food Industry: Even the Pickles Are Standardized

Needless to say, the food purveyed in fast-food restaurants is highly predictable. A short menu of simple foods helps ensure predictability. Hamburgers, fried chicken, pizza, tacos, French fries, soft drinks, shakes, and the like are all relatively easy to prepare and serve in a uniform fashion. Predictability in such products is made possible by the use of uniform raw ingredients, identical technologies for food preparation and cooking, similarity in the way the food is served, and identical packaging. As a trainer at Hamburger University puts it, "McDonald's has standards for everything down to the width of the pickle slices."[34]

Packaging is another important component of predictability in the fast-food restaurant. Despite the fast-food restaurants' best efforts, unpredictabilities can creep in because of the nature of the materials—the food might not be hot enough, the chicken might be gristly or tough, or there may be too few pieces of pepperoni on a particular slice of pizza. Whatever the (slight) unpredictabilities in the food, the packaging—containers for the burgers, bags for the small fries, cardboard boxes for the pizzas—can always be the same and imply that the food will be, too.

Predictable food also requires predictable ingredients. McDonald's has stringent guidelines on the nature (quality, size, shape, and so on) of the meat, chicken, fish, potatoes, and other ingredients purchased by each franchisee. The buns, for example, must be made of ordinary white bread from which all the chewy and nutritious elements of wheat, such as bran and germ, have been milled out. (One wit said of mass-produced white bread, "I thought they just blew up library paste with gas and sent it to the oven."[35]) Because buns otherwise might grow stale or moldy, preservatives are added to retard spoilage. Precut, uniform, frozen French fries rather than fresh potatoes are used.

The increasing use of frozen (or freeze-dried) foods addresses unpredictabilities related to the supply of raw materials. One of the reasons Ray Kroc eventually substituted frozen for fresh potatoes was that for several months a year, it was difficult to obtain the desired variety of potato. Freezing potatoes made them readily available year-round. In addition, the potato peelings at each McDonald's outlet often created a stench that was anathema to Kroc and the sanitized (sterile?) world he sought to create. Frozen, peeled, and precut French fries solved this problem as well.

The predictability of foods in a McDonaldized society has led to a disturbing fact: "Regional and ethnic distinctions are disappearing from American cooking. Food in one neighborhood, city, or state looks and tastes pretty much like food anywhere else. . . . Sophisticated processing and storage techniques, fast transport, and a creative variety of formulated convenience-food products have made it possible to ignore regional and seasonal differences in food production."[36]

Entertainment: Welcome to McMovieworld

The earlier discussion of *Psycho* brings to mind the fact that the movie industry, too, values predictability. *Psycho* was followed by several sequels (as well as a later [1998] shot-by-shot remake of the original). 2011 saw an unusually large number of sequels, including the fourth *Pirates of the Caribbean* (*On Stranger Tides*), *Scream 4, Spy Kids 4, Final Destination 5,* and Part 8 of the *Harry Potter* movies. The all-time leaders among sequels are the 12 *Friday the 13th* movies and the 23 *James Bond* movies. These predictable products generally attract a large audience, but they often succeed at the expense of movies based on new concepts, ideas, and characters.

The studios like sequels ("Welcome to McMovieworld")[37] because the same characters, actors, and basic plot lines can be used again and again. Furthermore, sequels are more likely to succeed at the box office than completely original movies; profits are therefore more predictable. Viewers presumably like sequels because they enjoy the comfort of encountering favorite characters played by familiar actors who find themselves in accustomed settings. Like a McDonald's meal, many sequels are not very good, but at least consumers know what they are getting.

Movies themselves seem to include increasingly predictable sequences and highly predictable endings. Dustin Hoffman contends that today's movie audiences would not accept the many flashbacks, fantasies, and dream sequences of his classic 1969 movie, *Midnight Cowboy.* He believes that this may be "emblematic of the whole culture . . . now people want to know what they're getting when they go to the movies."[38]

On television, the parallel to sequels is "copycatting" or producing sitcoms and comedies "that are so similar as to be indistinct."[39] For example, "They all gather in apartments and offices that tend to have heightened, overly colorful, casual-by-design look, and they exchange jokes that frequently depend on body parts or functions for their punch."[40] Among recent or current TV fare, *Seinfeld*, *Friends*, *Will & Grace*, *Everybody Loves Raymond*, *The Office*, *Parks and Recreation*, and *Two and a Half Men* come to mind. A variety of successful shows such as *CSI*, *NCIS*, and *Law & Order* have had several spinoffs (e.g., *NCIS Los Angeles*), and all follow a given formula. Successful TV shows such as *Survivor* and *American Idol* have spawned a number of shows that emulate the model they created. *Real Housewives of Orange County* has spawned, among others, *Real Housewives of New York City, Atlanta, Miami*, and *Beverly Hills*. "Like McDonald's, prime time wants you to know exactly what you'll get no matter where you are, emphasizing the comforts of predictability over nutrition."[41]

Another form of entertainment that aims to provide no surprises is the cruise, which is as oriented to predictability as it is to efficiency. A cruise originating from the United States will likely be made up of like-minded Americans. Cruise ship operators have turned travel into a highly predictable product by creating trips that allow minimal contact with the people, culture, and institutions of visited countries. This result creates a paradox: People go to considerable expense and effort to go to foreign countries, where they have as little contact as possible with native culture.[42]

American package tour agencies use American air carriers wherever possible or local transports that offer the amenities expected by American tourists (perhaps even air-conditioning, stereo, and bathroom). Tour guides are usually Americans or people who have spent time in America—at the very least, natives fluent in English who know all about the needs and interests of Americans. Restaurants visited on the tour are either American (perhaps associated with an American fast-food chain) or cater to the American palate. Hotels are also likely to be either American chains, such as the Sheraton and Hilton, or European hotels that have structured themselves to suit American tastes.[43] Each day offers a firm, often tight schedule, with little time for spontaneous activities. Tourists can take comfort from knowing exactly what they are going to do on a daily, even hourly, basis.

John Urry argues that the package tour has declined in popularity in recent years.[44] How do we reconcile this contention with the idea that McDonaldization is increasing? The answer lies in the fact that most societies have grown increasingly McDonaldized, with the result that there may be less need for McDonaldized tours. After all, since people traveling practically anywhere are likely to find McDonald's, Holiday Inn, Hard Rock Cafe, *USA TODAY,* and CNN, they may feel less need to be protected from unpredictabilities; many of them have already been eliminated.

Sports: There's Even a McStables

In tennis, the tiebreaker has made tennis matches more predictable. Prior to tiebreakers, to win a set, a player needed to win six games with a two-game margin over his or her opponent. If the opponent was never more than one game behind, the set could go on and on. Some memorable, interminable tennis matches produced scores on the order of 12 to 10. With limitations imposed by television and other mass media, the tennis establishment decided to institute the tiebreaker in many tournaments. If a set is deadlocked at six games each, a 12-point tiebreaker is played. The first player to get 7 points with a 2-point margin wins. A tiebreaker might go beyond 12 points (if the players are tied at 6 points each), but it rarely goes on nearly as long as close matches occasionally did.

An interesting example of predictability in a previously highly unpredictable area is in the rationalization of racehorse training. There is even a Circle McStables in Texas and a Three Bar McStables in Oklahoma. While these are one-of-a-kind operations, trainer D. Wayne Lukas has set up a string of stables around the United States that some have labeled "McStables." In the past, training stables were independent operations specific to a given track. Training procedures varied greatly from one racetrack to another and from one stable to another. Lukas, however, has been successful by establishing and supervising far-flung divisions of his stable. He said, "The barns are the same. The feeding program is the same. . . . There's never an adjustment necessary. . . . *It's the McDonald's principle*" (italics added).[45]

Minimizing Danger and Unpleasantness

The attraction of the shopping mall can be credited, at least in part, to its ability to make shopping more predictable. For example, "One kid who works here [at a mall] told me why he likes the mall. . . . It's because no matter what the weather is outside, it's always the same in here. He likes that. He doesn't want to know it's raining—it would depress him."[46] Those who wander through malls are also relatively free from the crime that might beset them in city streets. The lack of bad weather and the relative absence of crime point to another predictable aspect of shopping malls—they are always upbeat.

Avoiding crime is a key factor in the rise of so-called family fun or pay-to-play centers. (Often, children pay an entrance fee, although, in a cute gimmick, parents may be "free.") These centers offer ropes, padded "mountains," tubes, tunnels, giant blocks, trapezes, and so on. They have proven popular in urban areas because they provide a safe haven in crime-ridden cities.[47] Children are also seen as less likely to injure themselves in fun centers than in community playgrounds because of the nature of the equipment and the presence of staff supervisors. And there are safety checks to be sure that children do not leave with anyone but their parents. However, although fun centers are undoubtedly safer and less unpredictable, they have also been described as "antiseptic, climate-controlled, plastic world[s]."[48]

Modern amusement parks are in many ways much safer and more pleasant than their honky-tonk ancestors. The Disney organization quite clearly knew that to succeed, it had to overcome the unpredictability of old amusement parks. Disneyland and Walt Disney World take great pains to be sure that the visitor is not subject to any disorder. We have already seen how the garbage is whisked away so that people do not have to be disturbed by the sight of trash. Vendors do not sell peanuts, gum, or cotton candy because those items would make a mess underfoot. Visitors will not likely have their day disrupted by the sight of public drunkenness. Crime in the parks is virtually nonexistent. Disney offers a world of predictable, almost surreal, orderliness.

Few unanticipated things happen on any of the rides or in any of the attractions in contemporary theme parks. Of the Jungle Cruise ride at Disney World, a company publication says, "The Jungle Cruise is a favorite of armchair explorers, because it compresses weeks of safari travel into ten minutes [efficiency!] of fun, *without mosquitoes, monsoons, or misadventures*" (italics added).[49]

At one time, people went camping to escape the predictable routines of their daily lives. City dwellers fled their homes in search of nature, with little more than a tent and a sleeping bag. Little or nothing lay between the camper and the natural environment, leading to some unpredictable events. But that was the whole point. Campers might see a deer wander close to their campsite, perhaps even venture into it. Of course, they might also encounter the unexpected thunderstorm, tick bite, or snake, but these were accepted as an integral part of escaping one's routine activities.

Some people still camp this way; however, many others have sought to eliminate unpredictability from camping. Said the owner of one campground, "All they wanted [in the past] was a space in the woods and an outhouse. . . . But nowadays people aren't exactly roughing it."[50] Instead of simple tents, modern campers might venture forth in a recreational vehicle (RV) such as a Winnebago or take a trailer with an elaborate pop-up tent to protect them from the unexpected. Of course, "camping" in an RV also tends to reduce the likelihood of catching sight of wandering wildlife. Furthermore, the motorized camper carries within it all the elements one has at home—refrigerator, stove, television, portable DVD player, iPad, smartphone, and iPod base.

Camping technology not only has made for great predictability but has also changed modern campgrounds. Relatively few people now pitch their tents in the unpredictable wilderness; most find their way into rationalized campgrounds, even "country-club campgrounds," spearheaded by franchises such as Kampgrounds of America (KOA), the latter with more than 450 sites.[51] Said one camper relaxing in his air-conditioned, 32-foot trailer, "We've got everything right here. . . . It doesn't matter how hard it rains or how the wind blows."[52] Modern campgrounds are likely to be divided into sections—one for tents, another for RVs, each section broken into neat rows of usually tiny campsites. Hookups allow those with RVs to operate the various technologies encased within them. Campsite owners might also provide campers who are "roughing it" with amenities such as a well-stocked delicatessen, bathrooms and showers, heated swimming pools, a game room loaded with video games, a Laundromat, a TV room, a movie theater, and even entertainment such as bands or comedians.

There is certainly nothing wrong with wanting to be safe from harm. However, society as a whole has surrendered responsibility for providing safe environments to commercial interests. Because our city streets are often unsafe, people shop in malls. Because our playgrounds are sometimes unsafe (and greatly limited), children play in commercial "fun" centers. The problem is that people are therefore spending large amounts of leisure time in commercial environments eager to lead them into a life of consumption. If the larger society provided safe and attractive recreation centers for both adults and children, we would not be forced to spend so much of our lives, and do so many things, in commercial venues.

The irony is that, despite their claim to safety, McDonaldized locations, especially fast-food restaurants, seem particularly prone to crime and violence. Said the owner of a fast-food outlet, "Fast food for some reason is a target."[53] Perhaps the iron cage sometimes forces people to lash out in the setting that is its leading example.

Control: Human and Nonhuman Robots

The fourth dimension of McDonaldization is increased control of humans through the utilization of nonhuman technology. Technology includes not only machines and tools but also materials, skills, knowledge, rules, regulations, procedures, and techniques. Technologies thus encompass not only the obvious, such as robots and computers, but also the less obvious, such as the assembly line, bureaucratic rules, and manuals prescribing accepted procedures and techniques. A *human technology* (a screwdriver, for example) is controlled by people; a *nonhuman technology* (the order window at the drive-through, for instance) controls people.

The great source of uncertainty, unpredictability, and inefficiency in any rationalizing system is people—either those who work within it or those served by it. Hence, efforts to increase control are usually aimed at both employees and customers, although processes and products may also be the targets.

Organizations have historically gained control over people gradually, through increasingly effective technologies.[54] Eventually, they began reducing people's behavior to a series of machinelike actions. And once people were behaving like machines, they could be replaced with actual machines. The replacement of humans by machines is the ultimate stage in control over people; people can cause no more uncertainty and unpredictability because they are no longer involved, at least directly, in the process. As Erik Bryniolfsson and Andrew McAfee[55] point out in their 2011 book, *Race Against the Machine: How the Digital Revolution Is Accelerating Innovation, Driving Productivity, and Irreversibly Transforming Employment and the Economy,* large numbers of people in the age of accelerating digital technology are losing the race against technology. Computerized technology is now doing the work that at one time was the domain of people.

Control is not the only goal associated with nonhuman technologies. These technologies are created and implemented for many reasons, such as increased productivity, improved quality, and lower cost. However, this chapter is mainly concerned with the ways nonhuman technologies have increased control over employees and consumers in a McDonaldizing society.

Controlling Employees

Before the age of sophisticated nonhuman technologies, people were largely controlled by other people. In the workplace, owners and supervisors controlled subordinates directly, face to face. But such direct, personal control is difficult, costly, and likely to engender personal hostility. Subordinates might strike out at an immediate supervisor or an owner who exercises excessively tight control over their activities. Control through a technology is easier, less costly in the long run, and less likely to engender hostility toward supervisors and owners. Over time, control by people has thus shifted toward control by technologies.[56]

McDonald's controls employees by threatening to use, and ultimately using, technology to replace human workers. No matter how well they are programmed and controlled, workers can foul up the system's operation. A slow worker can make the preparation and delivery of a Big Mac inefficient. A worker who refuses to follow the rules might leave the pickles or special sauce off a hamburger, thereby making for unpredictability. And a distracted worker can put too few fries in the box, making an order of large fries seem skimpy. For these and other reasons, McDonald's and other fast-food restaurants have felt compelled to control and ultimately replace human beings with machines. Technology that increases control over workers helps McDonaldized systems assure customers that their products and services will be consistent.

The Fast-Food Industry:
From Human to Mechanical Robots

Fast-food restaurants have coped with problems of uncertainty by creating and instituting many nonhuman technologies. Among other things, they have done away with a cook, at least in the conventional sense. Of the cook, Jerry Newman says, "What a crock! Neither McDonald's nor Burger King had a 'cook.' Technology has all but eliminated this job."[57] Grilling a hamburger is so simple that anyone can do it with a bit of training. Even when more skill is required (as in the case of cooking an Arby's roast beef), the fast-food restaurant develops a routine involving a few simple procedures that almost anyone can follow. Cooking fast food is like a game of connect-the-dots or painting-by-numbers. Following prescribed steps eliminates most of the uncertainties of cooking.

Much of the food prepared at McDonaldized restaurants arrives preformed, precut, presliced, and "pre-prepared." All employees need to do, when necessary, is cook or often merely heat the food and pass it on to the customer. Instead of having employees cut lettuce at the local McDonald's restaurants in Great Britain, the following high-tech operation does it for them at a central location:

> The iceberg lettuce in the factory that supplies McDonald's is shredded at a rate of 1,000kg an hour by a 22-blade machine you wouldn't under any circumstances want to fall into. After that, its ordeal is far from over. It drops onto a conveyor belt which propels it a four metres per second into a flurolaser optical-sorting machine that cost the company a cool 350,000 pounds. The machine uses the latest in laser and digital-camera technology to analyse the colour and density of the lettuce shreds. It takes tens of thousands of scans a second to identify rogue chunks, grit and discolouration. When an anomaly is spotted, the air guns 50 cm further along the belt are primed. One-eighth of a second later, the jets blast the offending matter out of the flow. . . . Of course, lettuce is just the tip of the iceberg. Each constituent part of a Big Mac, from the burger to the sesame-seed bun, has a process of massive industrial scale behind it.[58]

At Taco Bell, workers used to spend hours cooking meat and shredding vegetables. The workers now simply drop bags of frozen, ready-cooked beef into boiling water. They have used preshredded lettuce for some time, and more recently preshredded cheese and prediced tomatoes have appeared.[59] The more that is done by nonhuman technologies before the food arrives at the restaurant, the less workers need to do and the less room they have to exercise their own judgment and skill.

McDonald's has gone further than most other chains in developing technologies to control employees. For example, its soft drink dispenser has a sensor that automatically shuts off the flow when the cup is full. Ray Kroc's dissatisfaction with the vagaries of human judgment led to the elimination of French fry machines controlled by humans and to the development of machines that ring or buzz when the fries are done or that automatically lift the French fry baskets out of the hot oil. When an employee controls the French fry machine, misjudgment may lead to undercooked, overcooked, or even burned fries. Kroc fretted over this problem: "It was amazing that we got them as uniform as we did, because each kid working the fry vats would have his own interpretation of the proper color and so forth."[60]

At the cash register, workers once had to look at a price list and punch the prices in by hand; hence, the wrong (even lower) amount could be rung up. Computer screens and computerized cash registers—point-of-sale registers—forestall that possibility.[61] All that the employees need do is press the image on the register that matches the item purchased; the machine then produces the correct price.

If the objective in a fast-food restaurant is to reduce employees to human robots, we should not be surprised by the spread of robots that prepare food. For example, a robot cooks hamburgers at one campus restaurant.[62]

Robots offer a number of advantages—lower cost, increased efficiency, fewer workers, no absenteeism, and a solution to the decreasing supply of teenagers needed to work at fast-food restaurants. The professor who came up with the idea for the robot that cooks hamburgers said, "Kitchens have not been looked at as factories, which they are. . . . Fast-food restaurants were the first to do that."[63]

Taco Bell developed "a computer-driven machine the size of a coffee table that . . . can make and seal in a plastic bag a perfect hot taco."[64] Another company worked on an automated drink dispenser that produced a soft drink in 15 seconds: "Orders are punched in at the cash register by a clerk. A computer sends the order to the dispenser to drop a cup, fill it with ice and appropriate soda, and place a lid on top. The cup is then moved by conveyor to the customer."[65]

These technologies have yet to be widely used, although Starbucks has its automated machines that grind beans and spit out espresso. When such technologies are refined and prove to be less expensive and more reliable than humans, fast-food restaurants will employ them widely.

Even managers are not immune from efforts to find nonhuman controls over them. There is a computerized system that, among other things, tells managers how many hamburgers or orders of French fries they will require at a given time (the lunch hour, for example). The computerized system takes away the need for managers to make such judgments and decisions.[66] Thus, "Burger production has become an exact science in which everything is regimented, every distance calculated and every dollop of ketchup monitored and tracked."[67]

Education: McChild Care Centers

An even more extreme version of the emphasis on nonhuman technology appears in the child-care equivalent of the fast-food restaurant, KinderCare, which was founded in 1969 and is now part of the Knowledge Learning Corporation. There are more than 1,600 KinderCare learning centers in the United States.[68] More than 300,000 children between the ages of 6 weeks and 12 years attend the centers.[69] KinderCare tends to hire short-term employees with little or no training in child care. What these employees do in the "classroom" is largely determined by an instruction book with a ready-made curriculum. Staff members open the manual to find activities spelled out in detail for each day. Clearly, a skilled, experienced, creative teacher is not the kind of person "McChild" care centers seek to hire; relatively untrained employees are more easily controlled by the nonhuman technology of the omnipresent "instruction book."

Another example of organizational control over teachers is the franchised Sylvan Learning Center, often thought of as the "McDonald's of Education."[70] (There are about 900 Sylvan Learning Centers worldwide, and they claim to have helped two million students.[71]) Sylvan Learning Centers are after-school centers for remedial education. The corporation "trains staff and tailors a McDonald's type uniformity, down to the U-shaped tables at which instructors work with their charges."[72] Through their training methods, rules, and technologies, for-profit systems such as the Sylvan Learning Center exert great control over their "teachers."

Health Care: Who's Deciding Our Fate?

As is the case with all rationalized systems, medicine has moved away from human and toward nonhuman technologies. The two most important examples are the growing importance of bureaucratic rules and controls and the growth of modern medical machinery. For example, the prospective payment and DRG (diagnosis-related group) systems—not physicians and their medical judgment—tend to determine how long a patient must be hospitalized. Similarly, the doctor operating alone out of a black bag with a few simple tools has become virtually a thing of the past. Instead, doctors serve as dispatchers, sending patients on to the appropriate machines and specialists. Computer programs can even diagnose illnesses.[73] Although it is unlikely that they will ever replace the physician, computers may one day be the initial, if not the prime, diagnostic agents. It is now even possible for people to get diagnoses, treatment, and prescriptions over the Internet with no face-to-face contact with a physician.

These and other developments in modern medicine demonstrate increasing external control over the medical profession by third-party payers, employing organizations, for-profit hospitals, health maintenance organizations (HMOs), the federal government (e.g., Medicare), and "McDoctor"-like organizations. Even in its heyday, the medical profession was not free of external control, but now the nature of the control is changing, and its degree and extent are increasing greatly. Instead of decisions being made by mostly autonomous doctors in private practice, doctors are more likely to conform to bureaucratic rules and regulations. In bureaucracies, employees are controlled by their superiors. Physicians' superiors are increasingly likely to be professional managers and not other doctors. Also, the existence of hugely expensive medical technologies often mandates that they be used. As the machines themselves grow more sophisticated, physicians come to understand them less and are therefore less able to control them. Instead, control shifts to the technologies as well as to the experts who create and handle them.

An excellent example of increasing external control over physicians (and other medical personnel) is called "pathways."[74] A pathway is a standardized series of steps prescribed for dealing with an array of medical problems. Involved are a series of "if-then" decision points—if a certain circumstance exists, the action to follow is prescribed. What physicians do in a variety of situations is determined by the pathway and not the individual physician. To put it in terms of this chapter, the pathway—a nonhuman technology—exerts external control over physicians.

Various terms have been used to describe pathways—standardization, "cookbook" medicine, a series of recipes, a neat package tied together with a bow, and so on—and all describe the rationalization of medical practice. The point is that there are prescribed courses of action under a wide range of circumstances. While doctors need not, indeed should not, follow a pathway at all times, they do so most of the time. A physician who spearheads the protocol movement says he grows concerned when physicians follow a pathway more than 92% of the time. While this leaves some leeway for physicians, it is clear that what they are supposed to do is predetermined in the vast majority of instances.

Let us take, for example, an asthma patient. In this case, the pathway says that, if the patient's temperature rises above 101°, a complete blood count is to be ordered. A chest X-ray is needed under certain circumstances—if it's the patient's initial wheezing episode or if there is chest pain, respiratory distress, or a fever of over 101°. And so it goes—a series of if-then steps prescribed for and controlling what physicians and other medical personnel do. While there are undoubted advantages associated with such pathways (e.g., lower likelihood of using procedures or medicines that have been shown not to work), they do tend to take decision making away from physicians. Continued reliance on such pathways is likely to adversely affect the ability of physicians to make independent decisions.

The Workplace: Do as I Say, Not as I Do

Most workplaces are bureaucracies that can be seen as large-scale nonhuman technologies. Their innumerable rules, regulations, guidelines, positions, lines of command, and hierarchies dictate what people do within the system and how they do it. The consummate bureaucrat thinks little about what is to be done; he or she simply follows the rules, deals with incoming work, and passes it on to its next stop in the system. Employees need do little more than fill out the required forms, these days most likely right on the computer screen.

At the lowest levels in the bureaucratic hierarchy ("blue-collar work"), scientific management clearly strove to limit or replace human technology. For instance, the "one best way" required workers to follow a series of predefined steps in a mindless fashion. More generally, Frederick Taylor believed that the most important part of the work world was not the employees but, rather, the organization that would plan, oversee, and control their work.

Although Taylor wanted all employees to be controlled by the organization, he accorded managers much more leeway than manual workers. The task of management was to study the knowledge and skills of workers and to record, tabulate, and ultimately reduce them to laws, rules, and even mathematical formulas. In other words, managers were to take a body of human skills, abilities, and knowledge and transform them into a set of nonhuman rules, regulations, and formulas. Once human skills were codified, the organization no longer needed skilled workers. Management would hire, train, and employ unskilled workers in accordance with a set of strict guidelines.

In effect, then, Taylor separated "head" work from "hand" work; prior to Taylor's day, the skilled worker had performed both. Taylor and his followers studied what was in the heads of those skilled workers, then translated that knowledge into simple, mindless routines that virtually anyone could learn and follow. Workers were thus left with little more than repetitive "hand" work. This principle remains at the base of the movement to replace human with nonhuman technology throughout our McDonaldizing society.

Behind Taylor's scientific management, and all other efforts at replacing human with nonhuman technology, lies the goal of being able to employ human beings with minimal intelligence and ability. In fact, Taylor sought to hire people who resembled animals, who had the mental makeup of oxen.[75] Not coincidentally, Henry Ford had a similar view of the kinds of people who were to work on his assembly lines: "The average worker, I am sorry to say, wants a job in which he does not have to think."[76] The kind of person sought out by Taylor was the same kind of person Ford thought would work well on the assembly line. In their view, such people would more likely submit to external technological control over their work and perhaps even crave such control.

Not surprisingly, a perspective similar to that held by Taylor and Ford can be attributed to other entrepreneurs: "The obvious irony is that the organizations built by W. Clement Stone [the founder of Combined Insurance] and Ray Kroc, both highly creative and innovative entrepreneurs, depend on the willingness of employees to follow detailed routines precisely."[77]

Many workplaces have come under the control of nonhuman technologies. In the supermarket, for example, the checker once had to read the prices marked on food products and enter them into the cash register. As with all human activities, however, the process was slow, with a chance of human error. To counter these problems, many supermarkets installed optical scanners, which "read" a code preprinted on each item. Each code number calls up a price already entered into the computer that controls the cash register. This nonhuman technology has thus reduced the number and sophistication of the tasks performed by the checker. Only the less-skilled tasks remain, such as physically scanning the food and bagging it. And even those tasks are being eliminated with the development of self-scanning and having consumers bag their own groceries. In other words, the work performed by the supermarket checker, when it hasn't been totally eliminated, has been "de-skilled"; that is, a decline has occurred in the amount of skill required for the job.

Control is exerted over the "phoneheads," or customer service representatives, who work for many companies. Those who handle reservations for the airlines (for example, United Airlines) must log every minute spent on the job and justify each moment away from the phone. Employees have to punch a "potty button" on the phone to let management know of their intentions. Supervisors sit in an elevated "tower" in the middle of the reservations floor, "observing like [prison] guards the movements of every operator in the room." They also monitor phone calls to make sure that employees say and do what they are supposed to. This control is part of a larger process of "omnipresent supervision increasingly taking hold in so many workplaces—not just airline reservations centers but customer service departments and data-processing businesses where computers make possible an exacting level of employee scrutiny."[78] No wonder customers often deal with representatives who behave like automatons. Said one employee of United Airlines, "My body became an extension of the computer terminal that I typed the reservations into. I came to feel emptied of self."[79]

Sometimes, telephone service representatives are literally prisoners. Prison inmates are now used in many states in this way, and the idea is currently on the legislative table in several more states. The attractions of prisoners are obvious—they work for very little pay and can be controlled to a far higher degree than even the "phoneheads" discussed above. Furthermore, they can be relied on to show up for work. As one manager put it, "I need people who are there every day."[80]

Following this logical progression, some companies now use computer callers instead of having people solicit us over the phone.[81] Computer voices are far more predictable and controllable than even the most rigidly controlled human operator, including prisoners. Indeed, in our increasingly McDonaldized society, I have had some of my most "interesting" conversations with such computer voices.

Of course, lower-level employees are not the only ones whose problem-solving skills are lost in the transition to more nonhuman technology. I have already mentioned the controls on professors and doctors. In addition, pilots flying the modern, computerized airplane (such as the Boeing 787 and the Airbus 380) are being controlled and, in the process, de-skilled. Instead of flying "by the seat of their pants" or occasionally using old-fashioned auto-pilots for simple maneuvers, modern pilots can "push a few buttons and lean back while the plane flies to its destination and lands on a predetermined runway." Said one FAA official, "We're taking more and more of these functions out of human control and giving them to machines." These airplanes are in many ways safer and more reliable than older, less technologically advanced models. Pilots, dependent on these technologies, however, may lose the ability to handle emergency situations creatively. The problem, said one airline manager, is that "I don't have computers that will do that [be creative]; I just don't."[82]

Controlling Customers

Employees are relatively easy to control because they rely on employers for their livelihood. Customers have much more freedom to bend the rules and go elsewhere if they don't like the situations in which they find themselves. Still, McDonaldized systems have developed and honed a number of methods for controlling customers.

The Fast-Food Industry: Get the Hell Out of There

Whether they go into a fast-food restaurant or use the drive-through window, customers enter a kind of conveyor system that moves them through the restaurant in the manner desired by the management. It is clearest in the case of the drive-through window (the energy for this conveyor comes from one's own automobile), but it is also true for those who enter the restaurant. Consumers know that they are supposed to line up, move to the counter, order their food, pay, carry the food to an available table, eat, gather up their debris, deposit it in the trash receptacle, and return to their cars.

Three mechanisms help control customers:[83]

1. Customers receive cues (for example, the presence of lots of trash receptacles, especially at the exits) that indicate what is expected of them.

2. A variety of structural constraints lead customers to behave in certain ways. For example, the drive-through window, as well as the written instructions on the menu marquee at the counter (and elsewhere), gives customers few, if any, alternatives.

3. Customers have internalized taken-for-granted norms and follow them when they enter a fast-food restaurant.

When my children were young, they admonished me after we finished our meal at McDonald's (I ate in fast-food restaurants in those days, before I "saw the light") for not cleaning up the debris and carting it to the trash can. My children were, in effect, serving as agents for McDonald's, teaching me the norms of behavior in such settings. I (and most others) have long-since internalized these norms, and I still dutifully follow them these days on the rare occasions that a lack of any other alternative (or the need for a reasonably clean restroom) forces me into a fast-food restaurant.

One goal of control in fast-food restaurants (Starbucks is an exception, at least to some degree; see Chapter 7) is to influence customers to spend their money and leave quickly. The restaurants need tables to be vacated rapidly so other diners will have a place to eat their food. A famous old chain of cafeterias, the Automat,[84] was partly undermined by people who occupied tables for hours on end. The Automat became a kind of social center, leaving less and less room for people to eat the meals they had purchased. The death-blow was struck when street people began to monopolize the Automat's tables.

Some fast-food restaurants employ security personnel to keep street people on the move or, in the suburbs, to prevent potentially rowdy teenagers from monopolizing tables or parking lots.

In some cases, fast-food restaurants have put up signs limiting a customer's stay in the restaurant (and even its parking lot), say, to 20 minutes. More generally, fast-food restaurants have structured themselves so that people do not need or want to linger over meals. Easily consumed finger foods make the meal itself a quick one. Some fast-food restaurants use chairs that make customers uncomfortable after about 20 minutes.[85] Much the same effect is produced by the colors used in the decor: "From the scarlet and yellow of the logo to the maroon of the uniform, everything clashes. It's designed to stop people from feeling so comfortable they might want to stay."[86]

Other Settings: It's Like Boot Camp

Grade schools have developed many ways to control students. Kindergarten has been described as an educational "boot camp."[87] Students are taught not only to obey authority but also to embrace the rationalized procedures of rote learning and objective testing. More important, spontaneity and creativity tend not to be rewarded and may even be discouraged, leading to what one expert calls "education for docility."[88] Those who conform to the rules are thought of as good students; those who don't are labeled bad students. As a general rule, the students who end up in college are the ones who have successfully submitted to the control mechanisms. Creative, independent students are often, from the educational system's point of view, "messy, expensive, and time-consuming."[89]

The clock and the lesson plan also exert control over students, especially in grade school and high school. Because of the "tyranny of the clock," a class must end at the sound of the bell, even if students are just about to comprehend something important. Because of the "tyranny of the lesson plan," a class must focus on what the plan requires for the day, no matter what the class (and perhaps the teacher) may find interesting. Imagine "a cluster of excited children examining a turtle with enormous fascination and intensity. Now children, put away the turtle, the teacher insists. We're going to have our science lesson. The lesson is on crabs."[90]

In the health care industry, the patient (along with the physician) is increasingly under the control of large, impersonal systems. For example, in many medical insurance programs, patients can no longer decide on their own to see a specialist. The patient must rather first see a primary care physician, who must decide whether a specialist is necessary. Because of the system's great pressure on the primary care physician to keep costs down, fewer patients visit specialists, and primary care physicians perform more functions formerly handled by specialists.

The supermarket scanners that control checkers also control customers. When prices were marked on all the products, customers could calculate roughly how much they were spending as they shopped. They could also check the price on each item to be sure that they were not being overcharged at the cash register. But with scanners, it is almost impossible for consumers to keep tabs on prices and on the cashiers.

Supermarkets also control shoppers with food placement. For example, supermarkets take pains to put the foods that children find attractive in places where youngsters can readily grab them (for example, low on the shelves). Also, what a market chooses to feature via special sale prices and strategic placement in the store profoundly affects what is purchased. Manufacturers and wholesalers battle one another for coveted display positions, such as at the front of the market or at the "endcaps" of aisles. Foods placed in these positions will likely sell far more than they would if they were relegated to their usual positions.

Malls also exert control over customers, especially children and young adults, who are programmed by the mass media to be avid consumers. Going to the mall can become a deeply ingrained habit. Some people are reduced to "zombies," shopping the malls hour after hour, weekend after weekend.[91] More specifically, the placement of food courts, escalators, and stairs forces customers to traverse corridors and pass attractive shop windows. Benches are situated so that consumers might be attracted to certain sites, even though they are seeking a brief respite from the labors of consumption. The strategic placement of shops, as well as goods within shops, leads people toward products in which they might not otherwise have been interested.

Controlling the Process and the Product

In a society undergoing McDonaldization, people are the greatest threat to predictability. Control over people can be enhanced by controlling processes and products, but control over processes and products also becomes valued in itself.

Food Production, Cooking, and Vending: It Cooks Itself

In the fast-food industry, companies have lengthy procedure manuals that exert considerable control over processes and products. For example, as a result of such procedures, Burger King Whoppers are cooked the same way at all restaurants in the chain, and the same is true for the McDonald's Quarter Pounders and their garnishes.[92]

Technologies designed to reduce uncertainties are also found throughout the manufacture of food. For example, the mass manufacturing of bread is not controlled by skilled bakers who lavish love and attention on a few loaves of bread at a time. Such skilled bakers cannot produce enough bread to supply the needs of our society. Furthermore, the bread they do produce can suffer from the uncertainties involved in having humans do the work. The bread may, for example, turn out to be too brown or too doughy. To increase productivity and eliminate these unpredictabilities, mass producers of bread have developed an automated system in which, as in all automated systems, humans play a minimal role rigidly controlled by the technology:

> The most advanced bakeries now resemble oil refineries. Flour, water, a score of additives, and huge amounts of yeast, sugar, and water are mixed into a broth that ferments for an hour. More flour is then added, and the dough is extruded into pans, allowed to rise for an hour, then moved through a tunnel oven. The loaves emerge after eighteen minutes, to be cooled, sliced, and wrapped.[93]

In one food industry after another, production processes in which humans play little more than planning and maintenance roles have replaced those dominated by skilled craftspeople. The warehousing and shipping of food has been similarly automated.

Further along in the food production process, other nonhuman technologies have affected how food is cooked. Technologies such as ovens with temperature probes "decide" for the cook when food is done. Many ovens, coffeemakers, and other appliances can turn themselves on and off. The instructions on all kinds of packaged foods dictate precisely how to prepare and cook the food. Premixed products, such as Mrs. Dash, eliminate the need for the cook to come up with creative combinations of seasonings. Even cookbooks were designed to take creativity away from the cook and control the process of cooking.

Some rather startling technological developments have occurred in the ways in which animals are raised for food. For instance, "aquaculture"[94] is growing dramatically, especially in China, because of the spiraling desire for seafood in an increasingly cholesterol-conscious world.[95] Instead of the old inefficient, unpredictable methods of harvesting fish—a lone angler casting a line or even boats catching tons of fish at a time in huge nets—we now have the much more predictable and efficient "farming" of seafood. More than half of the fresh salmon found in restaurants is now raised in huge sea cages off the coast of Norway. Almost all of the shrimp consumed in the United States is farmed and imported.

Sea farms offer several advantages. Most generally, aquaculture allows humans to exert far greater control over the vagaries that beset fish in their natural habitat, thus producing a more predictable supply. Various drugs and chemicals increase predictability in the amount and quality of seafood. Aquaculture also permits a more predictable and efficient harvest because the creatures are confined to a limited space. In addition, geneticists can manipulate them to produce seafood more efficiently. For example, it takes a standard halibut about 10 years to reach market size, but a new dwarf variety can reach the required size in only 3 years. Sea farms also allow for greater calculability—the greatest number of fish for the least expenditure of time, money, and energy.

Relatively small, family-run farms for raising other animals are being rapidly replaced by "factory farms."[96] The first animal to find its way into the factory farm was the chicken. Here is the way one observer describes a chicken "factory":

A broiler producer today gets a load of 10,000, 50,000, or even more day-old chicks from the hatcheries, and puts them straight into a long, windowless shed. . . . Inside the shed, every aspect of the birds' environment is controlled to make them grow faster on less feed. Food and water are fed automatically from hoppers suspended from the roof. The lighting is adjusted. . . . For instance, there may be bright light twenty-four hours a day for the first week or two, to encourage the chicks to gain [weight] quickly.[97]

Among its other advantages, such chicken farms allow one person to raise more than 50,000 chickens.

Raising chickens this way ensures control over all aspects of the business. For instance, the chickens' size and weight are more predictable than that of free-ranging chickens. "Harvesting" chickens confined in this way is also more efficient than catching chickens that roam over large areas. However, confining chickens in crowded quarters creates unpredictabilities, such as violence and even cannibalism. Farmers deal with these irrational "vices" in a variety of ways, such as dimming the lights as chickens approach full size and "debeaking" chickens so they cannot harm each other.

Some chickens are allowed to mature so they can be used for egg production. However, they receive much the same treatment as chickens raised for food. Hens are viewed as little more than "converting machines" that transform raw material (feed) into a finished product (eggs). Peter Singer describes the technology employed to control egg production:

> The cages are stacked in tiers, with food and water troughs running along the rows, filled automatically from a central supply. They have sloping wire floors. The slope . . . makes it more difficult for the birds to stand comfortably, but it causes the eggs to roll to the front of the cage where they can easily be collected . . . [and], in the more modern plants, carried by conveyor belt to a packing plant. . . . The excrement drops through [the wire floor] and can be allowed to pile up for many months until it is all removed in a single operation.[98]

This system obviously imposes great control over the production of eggs, leading to greater efficiency, a more predictable supply, and more uniform quality than the old chicken coop.

Other animals—pigs, lambs, steers, and calves especially—are raised similarly. To prevent calves' muscles from developing, which toughens the veal, they are immediately confined to tiny stalls where they cannot exercise. As they grow, they may not even be able to turn around. Being kept in stalls also prevents the calves from eating grass, which would cause their meat to lose its pale color; the stalls are kept free of straw, which, if eaten by the calves, would also darken the meat. "They are fed a totally liquid diet, based on nonfat milk powder with added vitamins, minerals, and growth-promoting drugs," says Peter Singer in his book *Animal Liberation*.[99] To make sure the calves take in the maximum amount of food, they are given no water, which forces them to keep drinking their liquid food. By rigidly controlling the size of the stall and the diet, veal producers can maximize two quantifiable objectives: the production of the largest amount of meat in the shortest possible time and the creation of the tenderest, whitest, and therefore most desirable veal.

Employment of a variety of technologies obviously leads to greater control over the process by which animals produce meat, thereby increasing the efficiency, calculability, and predictability of meat production. In addition, the technologies exert control over farm workers. Left to their own devices, ranchers might feed young steers too little or the wrong food or permit them too much exercise. In fact, in the rigidly controlled factory ranch, human ranch hands (and their unpredictabilities) are virtually eliminated.

The Ultimate Examples of Control: Birth and Death?

Not just fish, chickens, and calves are being McDonaldized through increasing control, but people are as well, especially in the processes of birth and death.

Controlling Conception: Even Granny Can Conceive

Conception is rapidly becoming McDonaldized, and increasing control is being exercised over the process. For example, the problem of male impotence[100] has been attacked by burgeoning impotence clinics, some of which have already expanded into chains,[101] and an increasingly wide array of technologies, including medicine (especially Viagra,[102] Cialis, and others) and mechanical devices. Many males are now better able to engage in intercourse and to play a role in pregnancies that otherwise might not have occurred.

Similarly, female infertility has been ameliorated by advances in the technologies associated with artificial (more precisely, "donor") insemination,[103] in vitro fertilization,[104] intracytoplasmic sperm injection,[105] various surgical and nonsurgical procedures associated with the Wurn technique,[106] at-home fertility kits,[107] and so on. Some fertility clinics have grown so confident that they offer a money-back guarantee if there is no live baby after three attempts.[108] For those women who still cannot become pregnant or carry to term, surrogate mothers can do the job.[109] Even postmenopausal women now have the chance of becoming pregnant ("granny pregnancies");[110] the oldest, thus far, is a 70-year-old Indian woman who gave birth to twins in 2008.[111] These developments and many others, such as ovulation-predictor home tests,[112] have made having a child far more predictable. Efficient, easy-to-use home pregnancy tests are also available to take the ambiguity out of determining whether a woman has become pregnant.

One of the great unpredictabilities tormenting some prospective parents is whether the baby will turn out to be a girl or a boy. Sex selection[113] clinics can be found in the United States, Canada, Australia, England, India, and Hong Kong, as part of what may eventually become a full-scale chain of "gender choice centers." The technology, developed in the early 1970s, is actually rather simple: Semen is filtered through albumin to separate sperm with male chromosomes from sperm with female chromosomes. The woman is then artificially inseminated with the desired sperm. The chances of creating a boy are 75%; a girl, 70%.[114] A new technique uses staining of sperm cells to determine which cells carry X (male) and Y (female) chromosomes. Artificial insemination or in vitro fertilization then mates the selected sperm with an egg. MicroSort technology offers a couple an 89.5% chance of creating a girl; the probabilities of creating a boy are 73.6%.[115] The goal is to achieve 100% accuracy in using "male" or "female" sperm to tailor the sex of the offspring to the needs and demands of the parents.

The increasing control over the process of conception delights some but horrifies others: "Being able to specify your child's sex in advance leads to nightmare visions of ordering babies with detailed specifications, like cars with automatic transmission or leather upholstery."[116] Said a medical ethicist, "Choosing a child like we choose a car is part of a consumerist mentality, the child becomes a 'product' rather than a full human being."[117] By turning a baby into just another "product" to be McDonaldized—engineered, manufactured, and commodified—people are in danger of dehumanizing the birth process.

Of course, we are just on the frontier of the McDonaldization of conception (and just about everything else). For example, the first cloned sheep, Dolly (now deceased), was created in Scotland in 1996, and other animals have since been cloned, opening the door to the possibility of cloning humans. Cloning involves creating identical copies of molecules, cells, or even entire organisms.[118] This conjures up an image of the engineering and mass production of a "cookie-cutter" race of people, all good-looking, athletic, intelligent, free of genetic defects, and so on—the ultimate in the control of conception. And a world in which everyone was the same would be a world in which they would be ready to accept a similar sameness in everything around them. Of course, this is a science fiction scenario, but the technology needed to take us down this road is already here!

Controlling Pregnancy: Choosing the Ideal Baby

Some parents wait until pregnancy is confirmed before worrying about the sex of their child, but then, amniocentesis can be used to determine whether a fetus is male or female. First used in 1968 for prenatal diagnosis, amniocentesis is a process whereby fluid is drawn from the amniotic sac, usually between the 14th and 18th weeks of pregnancy.[119] With amniocentesis, parents might choose to exert greater control over the process by aborting a pregnancy if the fetus is of the "wrong" sex. This technique is clearly far less efficient than pre-pregnancy sex selection because it occurs after conception. In fact, very few Americans (only about 5% in one study) say that they might use abortion as a method of sex selection.[120] However, amniocentesis does allow parents to know well in advance what the sex of their child will be.

Concern about a baby's sex pales in comparison to concerns about the possibility of genetic defects. In addition to amniocentesis, a variety of recently developed tests can be used to determine whether a fetus carries genetic defects such as cystic fibrosis, Down syndrome, Huntington's disease, hemophilia, Tay-Sachs disease, and sickle cell disease.[121] These newer tests include the following:

- Chorionic villus sampling (CVS): Generally done earlier than amniocentesis, between the 10th and 12th weeks of pregnancy, CVS involves taking a sample from the fingerlike structures projecting from the sac that later becomes the placenta. These structures have the same genetic makeup as the fetus.[122]
- Maternal serum alpha-fetoprotein (MSAFP) testing: This procedure is a simple blood test done in the 16th to 18th weeks of pregnancy. A high level of alpha-fetoprotein might indicate spina bifida; a low level might indicate Down syndrome.
- Ultrasound: This technology is derived from sonar and provides an image of the fetus by bouncing high-frequency energy off it. Ultrasound can reveal various genetic defects, as well as many other things (sex, gestational age, and so on).

The use of all these nonhuman technologies has increased dramatically in recent years, with some (ultrasound, MSAFP) already routine practices.[123] Many other technologies for testing fetuses are also available, and others will undoubtedly be created.

If one or more of these tests indicate the existence of a genetic defect, then abortion becomes an option. Parents who choose abortion are unwilling to inflict the pain and suffering of genetic abnormality or illness on the child and on the family. Eugenicists feel that it is not rational for a society to allow genetically disabled children to be born and to create whatever irrationalities will accompany their birth. From a cost-benefit point of view (calculability), abortion is less costly than supporting a child with serious physical or mental abnormalities or problems, sometimes for a number of years. Given such logic, it makes sense for society to use the nonhuman technologies now available to discover which fetuses should be permitted to survive and which should not. The ultimate step would be a societal ban on certain marriages and births, something China has considered, with the goal of reducing the number of sick or retarded children that burden the state.[124]

Efforts to predict and repair genetic anomalies are proceeding at a rapid rate. The Human Genome Project has constructed a map of the human genome's gene-containing regions.[125] When the project began, only about 100 human disease genes were known; today we know many more such genes.[126] This knowledge will allow scientists to develop new diagnostic tests and therapeutic methods. Identification of where each gene is and what each does will also extend the ability to test fetuses, children, and prospective mates for genetic diseases. Prospective parents who carry problematic genes may choose not to marry or not to procreate. Another possibility (and fear) is that, as the technology gets cheaper and becomes more widely available, people may be able to test themselves (we already have home pregnancy tests) and then make a decision to try a risky home abortion.[127] Overall, human mating and procreation will come to be increasingly affected and controlled by these new nonhuman technologies.

Controlling Childbirth: Birth as Pathology

McDonaldization and increasing control is also manifest in the process of giving birth. One measure is the decline of midwifery, a very human and personal practice. In 1900, midwives attended about half of American births, but by 1986, they attended only 4%.[128] Midwifery enjoyed a slight renaissance, however, because of the dehumanization and rationalization of modern childbirth practices,[129] and 6.5% of babies in the United States were being delivered by midwives in the late 20th century, and that increased by a third in the first decade of the 21st century.[130] (Worldwide, two thirds of all births involve midwives.[131]) When asked why they sought out midwives, women mention things such as the "callous and neglectful treatment by the hospital staff," "labor unnecessarily induced for the convenience of the doctor," and "unnecessary cesareans for the same reason."[132]

The use of midwives has also declined due to an increase in the control of the birth process by professional medicine,[133] especially obstetricians who are most likely to rationalize and dehumanize the birth process. Dr. Michelle Harrison, who served as a resident in obstetrics and gynecology, is but one physician willing to admit that hospital birth can be a "dehumanized process."[134]

The increasing control over childbirth is also demonstrated by the degree to which it has been bureaucratized. "Social childbirth," the traditional approach, once took place largely in the home, with female relatives and friends in attendance. Now, childbirth takes place almost totally in hospitals, "alone among strangers."[135] In 1900, less than 5% of U.S. births took place in hospitals; by 1940, hospital births totaled 55%, and by 1960, the process was all but complete, with nearly 100% of births occurring in hospitals.[136] In more recent years, hospital chains and birthing centers have emerged, modeled after my paradigm for the rationalization process—the fast-food restaurant.

Over the years, hospitals and the medical profession have developed many standard, routinized (McDonaldized) procedures for handling and controlling childbirth. One of the best known, created by Dr. Joseph De Lee, was widely followed throughout the first half of the 20th century. De Lee viewed childbirth as a disease (a "pathologic process"), and his procedures were to be followed even for low-risk births:[137]

1. The patient was placed in the lithotomy position, "lying supine with legs in air, bent and wide apart, supported by stirrups."[138]

2. The mother-to-be was sedated from the first stage of labor on.

3. An episiotomy[139] was performed to enlarge the area through which the baby must pass.

4. Forceps were used to make the delivery more efficient.

Describing this procedure, one woman wrote, "Women are herded like sheep through an obstetrical assembly line, are drugged and strapped on tables where their babies are forceps delivered."[140]

De Lee's standard practice includes not only control through nonhuman technology (the procedure itself, forceps, drugs, and an assembly-line approach) but most of the other elements of McDonaldization—efficiency, predictability, and the irrationality of turning the human delivery room into an inhuman baby factory. The calculability that it lacked was added later in the form of Emanuel Friedman's "Friedman Curve." This curve prescribed three rigid stages of labor. For example, the first stage was allocated exactly 8.6 hours, during which cervical dilation was to proceed from 2 to 4 centimeters.[141]

The moment babies come into the world, they are greeted by a calculable scoring system, the Apgar test. The babies receive scores of 0 to 2 on each of five factors (for example, heart rate, color), with 10 being the healthiest total score. Most babies have scores between 7 and 9 a minute after birth and scores of 8 to 10 after 5 minutes. Babies with scores of 0 to 3 are considered to be in very serious trouble. Dr. Harrison wonders why medical personnel don't ask about more subjective things, such as the infant's color, curiosity, and mood.[142]

The use of various nonhuman technologies for delivering babies has ebbed and flowed. The use of forceps, invented in 1588, reached a peak in the United States in the 1950s, when as many as 50% of all births involved their use. Forceps fell out of vogue, however, and by the 1980s, only about 15% of all births employed forceps, and their use continues to decline.[143] Many methods of drugging mothers-to-be have also been widely used. The electronic fetal monitor became popular in the 1970s. Today, ultrasound is a popular technology.

Another worrisome technology associated with childbirth is the scalpel. Many doctors routinely perform episiotomies during delivery so that the opening of the vagina does not tear or stretch unduly. Often done to enhance the pleasure of future sex partners and to ease the passage of the infant, episiotomies are quite debilitating and painful for the woman. Dr. Harrison expresses considerable doubt about episiotomies. "I want those obstetricians to stop cutting open women's vaginas. Childbirth is not a surgical procedure."[144]

The scalpel is also a key tool in cesarean sections. Birth, a perfectly human process, has come to be controlled by this technology (and those who wield it) in many cases.[145] The first modern "C-section" took place in 1882, but as late as 1970, only 5% of all births involved cesarean. Its use skyrocketed in the 1970s and 1980s, reaching 25% of all births in 1987 in what has been described as a "national epidemic."[146] By the mid-1990s, the practice had declined slightly, to 21%.[147] However, as of August 2002, 25% of births were once again by cesarean (and the rate in the United States was 30.3% by 2005, and it had risen to 32% by 2007),[148] first-time C-sections were at an all-time high of almost 17% in 2005, and the rate of vaginal births after a previous cesarean was down to 16.5%.[149] This latter occurred even though the American College of Obstetricians formally abandoned the time-honored idea, "once a cesarean, always a cesarean." That is, it no longer supports the view that, once a mother has a cesarean section, all succeeding births must be cesarean.

In addition, many people believe that cesareans are often performed unnecessarily. The first evidence for this belief is historical data: Why did we see a sudden need for so many more cesareans? Weren't cesareans just as necessary a few decades ago? The second clue regarding unnecessary cesareans is data indicating that private patients who can pay are more likely to get cesareans than those on Medicaid (which reimburses far less) and are twice as likely as indigent patients to get cesareans.[150] Are those with higher incomes really more likely to need cesareans than those with lower incomes?[151]

One explanation for the dramatic increase in cesareans is that they fit in well with the idea of substituting nonhuman for human technology, but they also mesh with the other elements of the McDonaldization of society:

- They are more predictable than the normal birth process, which can occur a few weeks (or even months) early or late. It is frequently noted that cesareans generally seem to be performed before 5:30 p.m. so that physicians can be home for dinner. Similarly, well-heeled women may choose a cesarean so that the unpredictabilities of natural childbirth do not interfere with careers or social demands.
- As a comparatively simple operation, the cesarean is more efficient than natural childbirth, which may involve many more unforeseen circumstances.
- Cesarean births are more calculable, normally involving no less than 20 minutes and no more than 45 minutes. The time required for a normal birth, especially a first birth, is far more variable.
- Irrationalities exist (see Chapter 5 for more on the irrationality of rationality), including the risks associated with surgery—anesthesia, hemorrhage, blood replacement. Compared with those who undergo a normal childbirth, women who have cesareans experience more physical problems and a longer period of recuperation, and the mortality rate can be as much as twice as high. There are also higher costs associated with cesareans. One study indicated that physicians' costs were 68% higher and hospital costs 92% higher for cesareans compared with natural childbirth.[152]
- Cesareans are dehumanizing because a natural human process is transformed, often unnecessarily, into a nonhuman or even inhuman process in which women endure a surgical procedure. At the minimum, many of those who have cesareans are denied unnecessarily the very human experience of vaginal birth. The wonders of childbirth are reduced to the routines of a minor surgical procedure.

Controlling the Process of Dying: Designer Deaths

We have now found ways to rationalize the dying process, giving us at least the illusion of control. Consider the increasing array of nonhuman technologies designed to keep people alive long after they would have expired had they lived at an earlier time in history. In fact, some beneficiaries of these technologies would not want to stay alive under those conditions (a clear irrationality). Unless the physicians are following an advance directive (a living will) that explicitly states "do not resuscitate" or "no heroic measures," people lose control over their own dying process. Family members, too, in the absence of such directives, must bow to the medical mandate to keep people alive as long as possible.

Computer systems may be used to assess a patient's chances of survival at any given point in the dying process—90%, 50%, 10%, and so on. The actions of medical personnel are likely to be influenced by such assessments.

Death has followed much the same path as birth; that is, the dying process has been moved out of the home and beyond the control of the dying and their families and into the hands of medical personnel and hospitals.[153] Physicians have gained a large measure of control over death, just as they won control over birth. Death, like birth, is increasingly likely to take place in the hospital. In 1900, only 20% of deaths took place in hospitals; by 2007, 32% of U.S. deaths occurred in hospitals, 20% occurred in nursing homes, and still another 38.5% occurred in hospices (2008 data).[154] The growth of hospital chains and chains of hospices, using principles derived from the fast-food restaurant, signals death's bureaucratization, rationalization, and even McDonaldization.

The McDonaldization of the dying process, as well as of birth, has spawned a series of counterreactions—efforts to cope with the excesses of rationalization. Advance directives and living wills tell hospitals and medical personnel what they may or may not do during the dying process. Suicide societies and books such as Derek Humphry's *Final Exit*[155] give people instructions on how to kill themselves. There is the growing interest in and acceptance of euthanasia,[156] most notably the work of "Dr. Death," Jack Kevorkian, whose goal was to return to people control over their own deaths. Finally, many people are choosing to die at home, and some are even opting to be buried there as well.[157]

However, these counterreactions themselves have elements of McDonaldization. For example, Dr. Kevorkian (who died in 2011) used a nonhuman technology, a "machine," to help people kill themselves. More generally, and strikingly, he was an advocate of a "rational policy" for the planning of death.[158] The rationalization of death is thus found even in the efforts to counter it.[159]

Overall, the future will bring with it an increasing number of nonhuman technologies with ever-greater ability to control people and processes. However, more and more people will lose the opportunity, and perhaps the ability, to think for themselves.

5

The Irrationality of Rationality

Traffic Jams on Those "Happy Trails"

McDonaldization has swept across the social landscape because it offers increased efficiency, predictability, calculability, and control. Despite these assets, as the preceding chapters have shown, McDonaldization has some serious drawbacks. Rational systems inevitably spawn irrationalities that limit, eventually compromise, and perhaps even undermine their rationality.

At the most general level, the irrationality of rationality is simply a label for many of the negative aspects of McDonaldization. More specifically, irrationality can be seen as the opposite of rationality. That is, McDonaldization can be viewed as leading to inefficiency, unpredictability, incalculability, and loss of control.[1] Other irrationalities of rationality to be discussed in this chapter include false friendliness, excessively high cost, health and environmental dangers, and homogenization. Irrationality also means that rational systems are disenchanted; they have lost their magic and mystery. Most important, rational systems are unreasonable systems that deny the humanity, the human reason, of the people who work within them or are served by them. In other words, rational systems are dehumanizing. Please note, therefore, that although the terms *rationality* and *reason* are often used interchangeably, in this discussion, they are antithetical phenomena[2]: Rational systems are often unreasonable.

Inefficiency: Long Lines at the Checkout

Rational systems, contrary to their promise, often end up being quite inefficient. For instance, in fast-food restaurants, long lines of people often form at the counters, or parades of cars idle in the drive-through lanes. What is purported to be an efficient way of obtaining a meal often turns out to be quite inefficient.

The problem of inefficiency at drive-throughs in the United States is, interestingly, among the greatest at McDonald's. While its goal is a maximum of a 90-second wait in the drive-through line, it averaged 152.5 seconds in 2004 and 167.9 seconds in 2005.[3] By 2009, the wait had increased further to 174.2 seconds. In fact, McDonald's ranked only seventh in 2009 in the average time waiting in its drive-throughs. The leader, with an average of 134 seconds per vehicle, was Wendy's.[4] Of course, to some degree, McDonald's is the victim of its own success, especially of the growth in its drive-through business. The paragon of efficiency, however, cannot hide behind its success and must deliver efficiency, even in the face of its burgeoning drive-through business. Furthermore, problems at the drive-through at McDonald's are not restricted to long waits—there are also many inaccuracies in their drive-through orders. In fact, in 2005, McDonald's ranked *last* among 25 fast-food chains in terms of accuracy. Said one customer, "McDonald's is the worst at getting things right. . . . McDonald's always gets at least one thing wrong."[5] And, it takes time to correct errors, further increasing the problem of inefficiency at McDonald's. Furthermore, time is money, and the increased time required to get an order right means higher costs and lower profits. In fact, one franchisee explicitly recognizes the irrational consequences of the rational emphasis on speed: "With continued emphasis on speed, accuracy suffers."[6]

While efficiency is a problem in the United States, it is a more serious problem elsewhere. A Hong Kong restaurant serves about 600,000 people a year (versus 400,000 in the United States). To handle long lines, 50 or more employees move along the lines taking orders with handheld computers. The orders are transmitted wirelessly to the kitchen.[7]

In the United States, a few McDonald's began experimenting with the use of handheld tablet PCs in mid-2011. An employee stands outside with the tablet taking orders from cars as soon as they get in line. The orders are transmitted wirelessly to the kitchen and are ready, at least theoretically, as soon as the car reaches the pick-up window. While this should speed things up, the initial experience was that the wait was "slow and frustrating."[8]

The fast-food restaurant is far from the only aspect of a McDonaldized society that exhibits inefficiency. Even the once-vaunted Japanese industry has its inefficiencies. Take the "just-in-time" system discussed in Chapter 2. Because this system often requires that parts be delivered several times a day, the streets and highways around a factory often became cluttered with trucks. Because of the heavy traffic, people were often late for work or for business appointments, resulting in lost productivity. But the irrationalities go beyond traffic jams and missed appointments. All these trucks used a great deal of fuel, very expensive in Japan, and contributed greatly to air pollution. The situation grew even worse when Japanese convenience stores, supermarkets, and department stores also began to use a just-in-time system, bringing even greater numbers of delivery trucks onto the streets.[9]

Here is the way columnist Richard Cohen describes another example of inefficiency in the McDonaldized world:

> Oh Lord, with each advance of the computer age, I was told I would benefit. But with each "benefit," I wind up doing more work. This is the ATM [automated teller machine] rule of life . . . I was told—nay promised—that I could avoid lines at the bank and make deposits or withdrawals any time of the day. Now, there are lines at the ATMs, the bank seems to take a percentage of whatever I withdraw or deposit, and, of course, I'm doing what tellers (remember them?) used to do. Probably, with the new phone, I'll have to climb telephone poles in the suburbs during ice storms.[10]

Cohen underscores at least three different irrationalities: (1) Rational systems are not less expensive, (2) they force people to do unpaid work, and, of most importance here, (3) they are often inefficient. It might be more efficient to deal with a human teller, either in the bank or at a drive-through window, than to wait in line at an ATM.

Similarly, preparing a meal might be more efficient at home than packing the family in the car, driving to McDonald's, loading up on food, and then driving home again. Meals cooked at home from scratch might not be more efficient, but certainly TV dinners and microwave meals are. They may even be more efficient than full-course meals picked up at the supermarket or Boston Market. Yet many people persist in the belief, fueled by propaganda from the fast-food restaurants, that eating there is more efficient than eating at home.

Although the forces of McDonaldization trumpet greater efficiency, they never tell us for whom the system is more efficient. Is it efficient for supermarket consumers who need only a loaf of bread and a carton of milk to wend their way past thousands of items they don't need? Is it efficient for consumers to push their own food over the supermarket scanner, swipe their own credit or debit cards, and then bag their groceries themselves? Is it efficient for people to pump their own gasoline? Is it efficient for them to push numerous combinations of telephone numbers before they hear a human voice? Most often, consumers find that such systems are *not* efficient for them. Most of the gains in efficiency go to those who are pushing rationalization.

Those at the top of an organization impose efficiencies not only on consumers but also on those who work at or near the bottom of the system: the assembly-line workers, the counter persons, the call-center staff. The owners, franchisees, and top managers want to control subordinates, but they want their own positions to be as free of rational constraints—as inefficient—as possible. Subordinates are to follow blindly the rules, regulations, and other structures of the rational system, while those in charge remain free to be "creative."

High Cost: Better Off at Home

The efficiency of McDonaldization (assuming it *is* efficient) does not ordinarily save consumers money. For example, some years ago, a small soda was shown to cost one franchise owner 11 cents, but it was sold for 85 cents.[11] A fast-food meal for a family of four might easily cost $28 (less if a few "Happy Meals" are in the mix) these days. Such a sum would go further if spent on ingredients for a home-cooked meal. For example, a meal for four (or even six) people, including a roast chicken, vegetables, salad, and milk, would cost about half that amount.[12] While oats, and real oatmeal, are very inexpensive, the relatively new oatmeal at McDonald's (see below) sells for $2.38, more than a double-cheeseburger. In fact, it is 10 times as expensive as the real thing.[13] A nutrition adviser points out that a Dollar Menu is "least economical" from the point of view of the nutrients obtained.[14]

As Cohen demonstrated with ATMs, people must often pay extra to deal with the inhumanity and inefficiency of rationalized systems. The great success of McDonaldized systems, the rush to extend them to ever-more sectors of society, and the fact that so many people want to get into such businesses indicate that these systems generate huge profits.

Bob Garfield noted the expense of McDonaldized activities in his article, "How I Spent (and Spent and Spent) My Disney Vacation."[15] Garfield took his family of four to Walt Disney World, which he found might more aptly be named "Expense World." The 5-day vacation cost $1,700 in 1991; admission to Disney World alone cost $551.30. And the prices keep going up. (Today, the cost of admission for 5 days for a family of four is about double that.) He calculates that, during the 5 days, they had less than 7 hours of "fun, fun, fun. That amounts to $261 c.p.f.h. (cost per fun hour)." Because most of his time in the Magic Kingdom was spent riding buses, "queuing up and shlepping from place to place, the 17 attractions we saw thrilled us for a grand total of 44 minutes."[16] Thus, what is thought to be an inexpensive family vacation turns out to be quite expensive.

False Friendliness: "Hi, George"

Because fast-food restaurants greatly restrict or even eliminate genuine fraternization, what workers and customers have left is either no human relationships or "false fraternization." Rule Number 17 for Burger King workers is, "Smile at all times."[17] The Roy Rogers employees who used to say "Happy trails" to me when I paid for my food really had no interest in what happened "on the trail." This phenomenon has been generalized to the many workers who say "Have a nice day" as customers depart. In fact, of course, they usually have no real interest in, or concern for, how the rest of a customer's day goes. Instead, in a polite and ritualized way, they are really saying, "Get lost," or move on so someone else can be served.

While they have declined dramatically in recent years as a result of e-mail and spam, we still receive computer-generated letters, or "junk mail" (of course, much more junk mail is now associated with our e-mail accounts). Great pains are sometimes taken to make a message seem personal.[18] (Similarly, I still get calls from telemarketers who start out by saying, "Hi, George.") In most cases, it is fairly obvious that a computer has generated the message from a database of names. These messages are full of the kind of false fraternization practiced by Roy Rogers workers. For example, they often adopt a friendly, personal tone designed to lead people to believe that the head of some business has fretted over the fact that they haven't, for example, shopped in his or her department store or used his or her credit card in the past few months. For example, a friend of mine received a letter from a franchise, The Lube Center, a few days after he had his car lubricated (note the use of the first name and the "deep" personal concern): "Dear Ken: We want to THANK YOU for choosing The Lube Center for all of your car's fluid needs. . . . We strongly recommend that you change your oil on a regular basis. . . . We will send you a little reminder card. . . . This will help *remind* you when your car is next due to be serviced. . . . We spend the time and energy to make sure that our employees are trained properly to give you the service that you deserve" (italics added).

Several years ago, I received the following letter from a congressman from Long Island, even though I live in Maryland. The fact that I had never met the congressman and knew nothing about him didn't prevent him from writing me a "personal" letter: "Dear George: It is hard to believe, but I am running for my NINTH term in Congress! When I think back over the 8,660 votes I've cast. . . . I realize how many battles *we've shared*. Please let me know that I can count on *you*" (italics added).

A *Washington Post* correspondent offers the following critique of false friendliness in junk mail:

> By dropping in people's names and little tidbits gleaned from databases hither and yon in their direct mail pitches, these marketing organizations are trying to create the illusion of intimacy. In reality, these technologies conspire to *corrupt and degrade intimacy.* They cheat, substituting the insertable fact for the genuine insight. These pitches end up with their own synthetic substitutes for the real thing.[19]

However false it may be, such junk mail is designed to exert control over customers by getting them to take desired courses of action.

Mention should also be made in this context of the false friendliness that defines greeting cards as well as e-cards.

Disenchantment: Where's the Magic?

One of Max Weber's most general theses is that, as a result of rationalization, the Western world has grown increasingly disenchanted.[20] The "magical elements of thought" that characterized less rationalized societies have been disappearing.[21] Thus, instead of a world dominated by enchantment, magic, and mystery, we have one in which everything seems clear, cut-and-dried, logical, and routine. As Schneider puts it, "Max Weber saw history as having departed a deeply enchanted past en route to a disenchanted future—a journey that would gradually strip the natural world both of its magical properties and of its capacity for meaning."[22] The process of rationalization leads, by definition, to the loss of a quality—enchantment—that was at one time very important to people. Although we undoubtedly have gained much from the rationalization of society in general, and from the rationalization of consumption settings in particular, we also have lost something of great, if hard to define, value. Consider how the dimensions of McDonaldization work against enchantment.

Efficient systems have no room for anything smacking of enchantment and systematically seek to root it out. Anything that is magical, mysterious, fantastic, dreamy, and so on is considered inefficient. Enchanted systems typically involve highly convoluted means to ends, and they may well have no obvious goals at all. Efficient systems do not permit such meanderings, and their designers and implementers will do whatever is necessary to eliminate them. The elimination of meanderings and aimlessness is one of the reasons that Weber saw rationalized systems as disenchanted systems.

Enchantment has far more to do with quality than with quantity. Magic, fantasies, dreams, and the like relate more to the inherent nature of an experience and the qualitative aspects of that experience rather than, for example, to the number of such experiences one has or the size of the setting in which they occur. An emphasis on producing and participating in a large number of experiences tends to diminish the magical quality of each of them. Put another way, it is difficult to mass produce magic, fantasy, and dreams. Such mass production may be common in the movies, but "true" enchantment is difficult, if not impossible, to produce in settings designed to deliver large quantities of goods and services frequently and over great geographic spaces. The mass production of such things is virtually guaranteed to undermine their enchanted qualities.

No characteristic of rationalization is more inimical to enchantment than predictability. Magical, fantastic, dreamlike experiences are almost by definition unpredictable. Nothing will destroy an enchanted experience more easily than having it become predictable or having it recur in the same way time after time.

Both control and the nonhuman technologies that produce control tend to be inimical to enchantment. As a general rule, fantasy, magic, and dreams cannot be subjected to external controls; indeed, autonomy is much of what gives them their enchanted quality. Fantastic experiences can go anywhere; anything can happen. Such unpredictability clearly is not possible in a tightly controlled environment. For some people, tight and total control could be a fantasy, but for many, it would be more of a nightmare. Much the same can be said of nonhuman technologies. Cold, mechanical systems are usually the antithesis of the dream worlds associated with enchantment. Again, some people have fantasies associated with nonhuman technologies, but they, too, tend to be more nightmarish than dreamlike.

As you can see, McDonaldization is related to, if not inextricably intertwined with, disenchantment. A world without magic and mystery is another irrational consequence of increasing rationalization.

Health and Environmental Hazards: A Day's Calories in One Fast-Food Meal

Progressive rationalization has threatened not only the fantasies but also the health, and perhaps the lives, of people. One example is the danger posed by the content of most fast food: a great deal of fat, cholesterol, salt, and sugar. Such meals are the last things Americans need, suffering as many of them do from obesity, high cholesterol levels, high blood pressure, and perhaps diabetes. In fact, there is much talk these days of an obesity epidemic (including children), and many observers place a lot of the blame on the fast-food industry—its foods and their contents—and its (continuing) emphasis on "supersizing" everything (even though they are now chary about using that term).[23]

The fast-food industry spent $4.2 billion in 2009 on advertising designed to convince people to consume its food. Furthermore, their processed food has an addictive quality. According to a former Food and Drug Administration commissioner, the industry created food that was "energy-dense, highly stimulating, and went down easy. They put it on every street corner and made it mobile, and they made it socially acceptable to eat anytime and anyplace. They created a food carnival . . . [we're] used to self-stimulation every 15 minutes."[24]

The negative impact of fast food on health is not restricted to the United States. The growth of fast-food restaurants, as well as the emphasis on ever-larger portions, is helping to lead to escalating health problems (e.g., diabetes) in various parts of the world, including the Far East in general and Vietnam in particular.[25] A comparative study of 380 regions in Ontario, Canada, showed that the regions with more fast-food services were likely to have higher rates of acute coronary syndrome and a higher mortality rate from coronary disease.[26]

Fast-food restaurants contribute to the development of various health problems later in life by helping create poor eating habits in children. By targeting children, fast-food restaurants are creating not only lifelong devotees of fast food but also people who are addicted to diets high in salt, sugar, and fat.[27] An interesting study discovered that the health of immigrant children deteriorates the longer they are in the United States, in large part because their diet begins to more closely resemble the junk-food diet of most American children.[28] In fact, Disney ended its long-term, cross-promotional relationship with McDonald's because of the growing concern about the link between fast food and childhood obesity.[29] A sociologist associated with the study of immigrant children stated, "The McDonaldization of the world is not necessarily progress when it comes to nutritious diets."[30]

Attacks against the fast-food industry's harmful effects on health have mounted over the years. Many of the franchises have been forced to respond by offering more and better salads, although the dressings for them are often loaded with salt and fat. They have also been forced to list nutritional information for all products in their stores and online. Still, most consumers never consult these lists and continue to order the typical McDonald's meal of a Big Mac, large fries, and a shake, which totals more than 1,500 calories, few of them of great nutritional value. The trend toward larger and larger portions has only increased the problem (McDonald's Big Breakfast with hotcakes and a large biscuit has 1,150 calories). The substitution of a chocolate triple shake (with 1,160 calories) for a regular shake in that McDonald's meal raises the total to 2,290 calories.[31] Burger King's Double Whopper with cheese alone has 976 calories (and 6 grams of fat).[32] Recommended calorie intake per day for women is less than 2,000 calories and just above 2,500 calories for men. Thus, just the typical McDonald's meal with the triple shake exceeds the recommended daily calorie intake for women and is close to the recommended amount of daily calories for men.

McDonald's new oatmeal is called a "bowl full of wholesome."[33] In itself, unadulterated oatmeal is a healthy food, but predictably McDonald's has done "everything it can to turn oatmeal into yet another bad choice."[34] One observer described the ingredients as "oats, sugar, sweetened dried fruit, cream and 11 weird ingredients you would never keep in your kitchen."[35] It contains more sugar than a Snickers candy bar. The observer wonders why McDonald's would take a healthy food like oats "and turn it into expensive junk food? Why create a hideous concoction of 21 ingredients, many of them chemical and/or unnecessary?"[36]

Although McDonald's (and others) has responded to its critics by changing its menu a bit and by offering publicly available nutritional facts about its products, it also reacted, predictably, with an ad campaign. In one ad, Ronald McDonald is seen as a "sports dude," juggling vegetables and dodging strawberries, without a burger in view. Experts, however, see such ads for what they are—publicity— while McDonald's continues to push fatty, high-calorie foods and huge portion sizes. Given the epidemic of childhood obesity, the huge sums spent on the marketing of such foods to children have to be of particular concern.[37]

McDonaldization poses even more immediate health threats. Regina Schrambling links outbreaks of diseases such as *Salmonella* to the rationalization of chicken production: "Salmonella proliferated in the poultry industry only after . . . Americans decided they wanted a chicken in every pot every night. But birds aren't like cars: you can't just speed up the factory line to meet demand. . . . Birds that are rushed to fryer size, then killed, gutted, and plucked at high speed in vast quantities are not going to be the cleanest food in the supermarket.[38] Schrambling also associates *Salmonella* with the rationalized production of eggs, fruit, and vegetables.[39]

Outbreaks of *Escherichia coli* infections have also been increasing in recent years, and the fast-food industry has taken note of this fact. Indeed, the first reported outbreak in the United States was traced to McDonald's in 1982. In 1997, Hudson Foods, a meatpacking company that supplied meat to McDonald's and Burger King, among others, was forced out of business because an outbreak of *E. coli* was traced to its frozen hamburgers.[40] Taco Bell experienced an outbreak of *E. coli* in 2006 that was traced to contaminated lettuce. Hamburger is a particular culprit because *E. coli* can be passed from steer to steer, and ultimately the hamburger from many steers, some of it infected, is mixed together. That meat is then turned into patties and frozen, and those frozen patties are distributed widely. The fast-food industry did respond to the danger of *E. coli* by cooking its hamburgers at a higher temperature to kill the bacterium, but *E. coli* is finding its way into an ever-larger number of highly McDonaldized foods (e.g., bagged salad and spinach).[41] *E. coli* remains a great concern today, especially in McDonaldized foods of all types.[42]

The fast-food industry has run afoul not only of nutritionists and epidemiologists but also of environmentalists; McDonald's and McDonaldization have produced a wide array of adverse effects on the environment. For example, the fast-food industry is directly linked to an enormous increase in meat production (projected to grow from 275 million tons in 2007 to 465 million tons in 2050[43]) and consumption. This increase in meat production is associated with a number of environmental problems such as land degradation, climate change, water and air pollution, water shortage, and a decline in biodiversity.[44] Large-scale hog farms, for example, produce a huge amount of manure that ultimately finds its way into our waterways and then our drinking water; people have been made ill and women have had miscarriages as a result of drinking water contaminated in this way.[45] The dosing of factory-farmed animals with antibiotics may lead to bacteria that are resistant to antibiotics, thereby putting people at risk.[46] Aquaculture creates a similar set of environmental problems and health risks to humans.[47]

Another adverse environmental effect stems from the need to grow uniform potatoes from which to create predictable French fries. The huge farms of the Pacific Northwest that now produce such potatoes rely on the extensive use of chemicals. In addition, the need to produce a perfect fry means that much of the potato is wasted, with the remnants either fed to cattle or used for fertilizer. The underground water supply in the area is now showing high levels of nitrates, which may be traceable to the fertilizer and to animal wastes.[48]

The fast-food industry produces an enormous amount of trash, some of which is non-biodegradable. The litter from fast-food meals is a public eyesore. Innumerable square miles of forest are sacrificed to provide the paper needed each year by McDonald's alone.[49] Whole forests are being devoured by the fast-food industry. For a time, paper containers were replaced by Styrofoam and other products. However, the current trend is back to paper (and other biodegradable) products; Styrofoam, virtually indestructible, piles up in landfills, creating mountains of waste that endure there for years, if not forever. Overall, despite various efforts to deal with its worst abuses, the fast-food industry contributes to global warming, destruction of the ozone layer, depletion of natural resources, and destruction of natural habitats.

Of course, the above merely scratches the surface of the ecological problems associated with the McDonaldization of the fast-food industry. To take another specific example, great inefficiency and huge environmental effects are associated with the care and feeding of immense herds of cattle. That is, it would be far more efficient for us to consume the grain ourselves than it is to consume the much smaller amount of beef derived from cattle that are grain-fed. More generally, this is all part of a fast-paced, highly mobile, and vast energy-consuming way of life that is having untold negative effects on the ecology of the world. While it is impossible to calculate exactly what they contribute to the problem, there is no question that the fast-food industry and McDonaldized systems in general are significant causes of a number of potential global calamities.[50]

The automobile assembly line has been extraordinarily successful in churning out millions of cars a year. But all those cars have wreaked havoc on the environment. Their emissions pollute the air, soil, and water; an ever-expanding system of highways and roads have scarred the countryside; and we must not forget the thousands of people killed and the far greater number injured each year in traffic accidents. It was the widespread use of the automobile that helped lead to the fast-food industry, and the nature of fast-food restaurants (their locations and their drive-through windows) encourages ever-greater use of the automobile.

Homogenization: It's No Different in Paris

Another irrational effect of McDonaldization is increased homogenization. Anywhere you go in the United States and, increasingly, throughout the world, you are likely to find the same products offered in the same way. The expansion of franchising across the United States means that people find little difference between regions and between cities.[51] On a global scale, travelers are finding more familiarity and less diversity. Exotic settings are increasingly likely sites for American fast-food chains and other McDonaldized settings.

Furthermore, in many nations, restaurant owners are applying the McDonald's model to native cuisine. In Paris, tourists may be shocked by the number of American fast-food restaurants but even more shocked by the incredible spread of indigenous forms, such as the fast-food croissanterie. One would have thought that the French, who seem to consider the croissant a sacred object, would resist rationalizing its manufacture and sale, but that is just what has happened.[52] The spread of such outlets throughout Paris indicates that many Parisians are willing to sacrifice quality for speed and efficiency. And, you may ask, if the Parisian croissant can be tamed and transformed into a fast-food success, what food is safe?

The spread of American and indigenous fast food causes less and less diversity from one setting to another. In the process, the human craving for new and diverse experiences is being limited, if not progressively destroyed. It is being supplanted by the desire for uniformity and predictability.

In general, McDonaldized institutions have been notably unsuccessful in creating new and different products. Please recall Ray Kroc's failures in this realm, notably the Hulaburger. Such systems excel instead at selling familiar products and services in shiny new settings or packages that can be easily replicated. For instance, the fast-food restaurant wraps that prosaic hamburger in bright packages and sells it in a carnival-like atmosphere that differs little from one locale to another. This point extends to many other manifestations of McDonaldization. For example, Jiffy Lube and its imitators sell people nothing more than the same old oil change and lube job.

Just as the franchises are leveling differences among goods and services, online consumption sites and online and mail-order catalogs are eliminating temporal and seasonal differences. When columnist Ellen Goodman received her Christmas catalog at the beginning of the fall, she offered this critique: "The creation of one national mail-order market has produced catalogues without the slightest respect for any season or region. Their holidays are now harvested, transported and chemically ripened on the way to your home. . . . I refuse to fast forward through the fall."[53]

Dehumanization: Getting Hosed at "Trough and Brew"

The main reason to think of McDonaldization as irrational, and ultimately unreasonable, is that it tends to be dehumanizing. For example, the fast-food industry offers its employees "McJobs."[54] One "McTask" at McDonald's is known as HBO—"Hand Bag Out."[55] As a Burger King worker has noted, "A moron could learn this job, it's so easy," and "Any trained monkey could do this job."[56] Workers can use only a small portion of their skills and abilities. The minimal skill demands of the fast-food restaurant are irrational.

From the employee's perspective, McJobs are irrational because they don't offer much in the way of either satisfaction or stability. Employees are seldom allowed to use anything approaching all their skills and are not allowed to be creative on the job. The result is a high level of resentment, job dissatisfaction, alienation, absenteeism, and turnover.[57] In fact, the fast-food industry has the highest turnover rate—approximately 300% a year[58]—of any industry in the United States. That means that the average fast-food worker lasts only about 4 months; the entire workforce of the fast-food industry turns over approximately three times a year.

Although the simple and repetitive nature of the jobs makes it relatively easy to replace workers who leave, such a high turnover rate is undesirable from the organization's perspective, as well as the employee's. It would clearly be better to keep employees longer. The costs involved in turnover, such as hiring and training, greatly increase with extraordinarily high turnover rates. In addition, failure to use employees' skills is irrational for the organization, for it could obtain much more from its employees for the money (however negligible) it pays them.

The automobile assembly line is well known for the way it dehumanizes life on a day-to-day basis for those who work on it. Although Henry Ford felt, as we saw earlier, that he personally could not do the kind of repetitive work required on the assembly line, he believed that most people, with their limited mental abilities and aspirations, could adjust to it quite well. Ford said, "I have not been able to discover that repetitive labour injures a man in any way. . . . The most thorough research has not brought out a single case of a man's mind being twisted or deadened by the work."[59] Objective evidence of the destructiveness of the assembly line, however, is found in the high rates of absenteeism, tardiness, and turnover among employees. More generally, most people seem to find assembly-line work highly alienating. Here is the way one worker describes it: "I stand in one spot, about a two—or three—[foot] area, all night. The only time a person stops is when the line stops. We do about thirty-two jobs per car, per unit, forty-eight units an hour, eight hours a day. Thirty-two times forty-eight times eight. Figure it out, that's how many times I push that button."[60]

Another worker offers a similar view: "What's there to say? A car comes, I weld it; a car comes, I weld it; a car comes, I weld it. One hundred and one times an hour." Others get quite sarcastic about the nature of the work: "There's a lot of variety in the paint shop. . . . You clip on the color hose, bleed out the color and squirt. Clip, bleed, squirt; clip, bleed, squirt, yawn; clip, bleed, squirt, scratch your nose."[61] Another assembly-line worker sums up the dehumanization he feels: "Sometimes I felt just like a robot. You push a button and you go this way. You become a mechanical nut."[62]

Alienation affects not only those who work on the automobile assembly line but also people in the wide range of settings built, at least in part, on the principles of the assembly line.[63] In our McDonaldized society, the assembly line has implications for many of us and for many different settings. The demands in the meatpacking industry (which is heavily dependent on the business provided by fast-food restaurants) are responsible, at least in part, for increasing dehumanization—inhuman work in inhumane conditions. Workers reduced to fast-moving cogs in the assembly-line killing and butchering of animals. They are forced to perform repetitive and physically demanding tasks on animals that may, at least initially, not even be dead. They are often covered in, and forced to stand in pools of, blood. They wield very sharp knives at great speed in close proximity to other workers. The result is an extraordinarily high injury (and even death) rate, although many injuries go unreported out of fear of being fired for being injured and unable to perform at peak levels. Because they are often undocumented immigrants, workers are almost totally at the whim of a management free to hire and fire them at will. Management is also able to ignore the horrid working conditions confronted by these powerless employees or able to make those conditions even more horrific.[64]

The fast-food restaurant also dehumanizes consumers. By eating on a sort of assembly line, diners are reduced to automatons rushing through a meal with little gratification derived from the dining experience or from the food itself. The best that can usually be said is that the meal is efficient and is over quickly. Typical diners at a McDonald's in Washington, D.C., are described as "slouching toward a quick and forgettable meal."[65]

Some customers might even feel as if they are being fed like livestock on an assembly line. This point was made on TV a number of years ago in a *Saturday Night Live* parody of a small fast-food chain called "Burger and Brew." In the skit, some young executives learn that a new fast-food restaurant called "Trough and Brew" has opened, and they decide to try it for lunch. When they enter the restaurant, bibs are tied around their necks. They then discover what resembles a pig trough filled with chili and periodically refilled by a waitress scooping new supplies from a bucket. The customers bend over, stick their heads into the trough, and lap up the chili as they move along the trough, presumably making "high-level business decisions" en route. Every so often they come up for air and lap some beer from the communal "brew basin." After they have finished their "meal," they pay their bills "by the head." Since their faces are smeared with chili, they are literally "hosed off" before they leave the restaurant. The young executives are last seen being herded out of the restaurant, which is closing for a half hour so that it can be "hosed down." *Saturday Night Live* was clearly ridiculing the fact that fast-food restaurants tend to treat their customers like lower animals.

Customers are also dehumanized by scripts and other efforts to make interactions uniform. "Uniformity is incompatible when human interactions are involved. Human interactions that are mass-produced may strike consumers as dehumanizing if the routinization is obvious or manipulative if it is not."[66] In other words, dehumanization occurs when prefabricated interactions take the place of authentic human relationships. Bob Garfield's critique of Walt Disney World provides another example of dehumanized customers: "I actually believed there was real fun and real imagination in store only to be confronted with an extruded, injection-molded, civil-engineered brand of fantasy, which is to say: no fantasy at all. . . . From the network of chutes and corrals channeling people into attractions, to *the chillingly programmed Stepford Wives demeanor of the employees,* to the compulsively litter-free grounds, to the generalized North Korean model Socialist Society sense of totalitarian order, to the utterly passive nature of the entertainment itself, Disney turns out to be the very antithesis of fantasy" (italics added).[67] Thus, instead of being creative and imaginative, Disney World turns out to be an uncreative, unimaginative, and ultimately inhuman experience.

Fast-food restaurants and other McDonaldized settings also minimize contact among human beings. The relationships between employees and customers are fleeting at best. Because employees typically work part-time and stay only a few months, even regular customers rarely develop personal relationships with them. All but gone are the days when one got to know well a waitress at a diner or the short-order cook at a local "greasy spoon." There are fewer and fewer places where an employee knows who you are and knows what you are likely to order. Fast being overwhelmed are what Ray Oldenburg calls "great good places," such as local cafés and taverns.[68]

Contact time between workers and customers at the fast-food restaurant is also very short. It takes little time at the counter to order, receive the food, and pay for it. Both employees and customers are likely to feel rushed and to want to move customers on to their meal and employees to the next order.[69] There is virtually no time for customer and counter person to interact. This is even truer of the drive-through window where, thanks to the speedy service and the physical barriers, the server is even more distant.

Other potential relationships in fast-food restaurants are also limited greatly. Because employees remain on the job for only a few months, satisfying personal relationships among employees are unlikely to develop. More permanent employment helps foster long-term relationships on the job, and workers with more job stability are likely to get together after work hours and on weekends. The temporary and part-time character of jobs in fast-food restaurants, and other McDonaldized settings, largely eliminates the possibility of such personal relationships among employees.

Relationships among fast-food customers are largely curtailed as well. Although some McDonald's ads would have people believe otherwise, gone for the most part are the days when people met in the diner or cafeteria for coffee or a meal and lingered to socialize. Fast-food restaurants clearly do not encourage such socializing. The exception seems to be Starbucks, but as we will see in Chapter 7, this is more myth than reality.

Family: The Kitchen as Filling Station

Fast-food restaurants also negatively affect the family, especially the so-called family meal.[70] The fast-food restaurant is not conducive to a long, leisurely, conversation-filled dinnertime. Furthermore, because of the fast-food restaurant, teens are better able to go out and eat with their friends, leaving the rest of the family to eat somewhere else or at another time. Of course, the drive-through window only serves to reduce further the possibility of a family meal. The family that gobbles its food while driving on to its next stop can hardly enjoy "quality time."

Here is the way one journalist describes what is happening to the family meal:

Do families who eat their suppers at the Colonel's, swinging on plastic seats, or however the restaurant is arranged, say grace before picking up a crispy brown chicken leg? Does dad ask junior what he did today as he remembers he forgot the piccalilli and trots through the crowds over to the counter to get some? Does mom find the atmosphere conducive to asking little Mildred about the problems she was having with third conjugation French verbs, or would it matter since otherwise the family might have been at home chomping down precooked frozen food, warmed in the microwave oven and watching [television]?[71]

There is much talk these days about the disintegration of the family, and the fast-food restaurant may well be a crucial contributor to that disintegration. Conversely, the decline of the family creates ready-made customers for fast-food restaurants.

In fact, dinners at home may now be not much different from meals at the fast-food restaurant. Families long ago tended to stop having lunch and breakfast together. Today, the family dinner is following the same route. Even at home, the meal is probably not what it once was. Following the fast-food model, people have ever more options to "graze," "refuel," nibble on this, or snack on that rather than sit down at a formal meal. Also, because it may seem inefficient to do nothing but just eat, families are likely to watch television, play computer games, or text or tweet while they are eating; they might even split up and, plate in hand, head for their own computers.[72] The din, to say nothing of the lure, of dinnertime TV programs such as *Wheel of Fortune,* the "buzzes" and "beeps" associated with smartphones, and the distraction associated with sending and receiving text messages are likely to make it difficult for family members to interact with one another. We need to decide whether we can afford to lose the primary ritual of the communal meal: "If it is lost to us, we shall have to invent new ways to be a family. It is worth considering whether the shared joy that food can provide is worth giving up."[73]

Beyond the computer and the smartphone, a key technology in the destruction of the family meal has been the microwave oven and the vast array of microwavable foods it helped generate.[74] Some time ago, a *Wall Street Journal* poll indicated that Americans consider the microwave their favorite household product (that would undoubtedly be replaced by the smartphone and the laptop computer today). Said one consumer researcher, "It has made even fast-food restaurants not seem fast because at home you don't have to wait in line." As a general rule, consumers demand meals that take no more than 10 minutes to microwave, whereas in the past, people were more often willing to spend a half hour or even an hour cooking dinner. This emphasis on speed has, of course, brought with it lower quality, but people do not seem to mind this loss: "We're just not as critical of food as we used to be."[75]

The speed of microwave cooking and the wide variety of microwavable foods make it possible for family members to eat at different times and places. With microwaveable products such as Hormel "Completes" Kids Microwave Meals and "Kid Cuisine" (the latter has similar products in frozen food), even children can "zap" their own meals. As a result, "Those qualities of the family meal, the ones that imparted feelings of security and well-being, might be lost forever when food is 'zapped' or 'nuked' instead of cooked."[76]

The advances in microwave cooking continue. On some foods, plastic strips turn blue when the food is done. The industry has even promised strips that communicate cooking information directly to the microwave oven. "With cooking reduced to pushing a button, the kitchen may wind up as a sort of filling station. Family members will pull in, push a few buttons, fill up and leave. To clean up, all we need do is throw away plastic plates."[77]

The family meal is not the only aspect of family life threatened by McDonaldization. For example, busy and exhausted parents are being advised that, instead of reading to their children at night, they should have them listen to audiotapes.[78]

Higher Education: McLectures and McColleges

The modern university has, in various ways, become a highly irrational place. The impact of McDonaldization is clear, for example, in the way that students increasingly relate to professors as if they were workers in the fast-food industry. If the "service" in class is not up to their standards, students feel free to complain and even behave abusively toward their professors. Both students and faculty members are put off by schools' factory-like atmosphere. They may feel like automatons processed by the bureaucracy and computers, or feel like cattle run through a meat-processing plant. In other words, education in such settings can be a dehumanizing experience.

The masses of students, large and impersonal dorms, and huge lecture classes make getting to know other students difficult. The large lecture classes, constrained tightly by the clock, also make it virtually impossible to know professors personally; at best, students might get to know a graduate assistant teaching a discussion section. Grades (and students are obsessed by this quantifiable measure of education) might be derived from a series of machine-graded, multiple-choice exams and posted on Blackboard. In sum, students may feel like little more than objects into which knowledge is poured as they move along an information-providing and degree-granting educational assembly line.

Of course, technological advances are leading to even greater irrationalities in education. The minimal contact between teacher and student is being limited further by advances such as Internet education and online instruction,[79] educational television, closed-circuit television,[80] distance learning, computerized instruction, and teaching machines. With Internet education, we have seen the ultimate step in the dehumanization of education: the elimination of a human teacher and of human interaction between teacher and student. Said one historian, "Taking a course online, by yourself, is not the same as being in a classroom with a professor who can respond to you, present different viewpoints and push you to work a problem."[81] Where human teachers do remain, they are less likely to be tenured professors and more likely to be part-time employees ("McLecturers" at "McColleges"),[82] who are apt to be treated as disposable service workers by both the university and students.

Health Care: You're Just a Number

For the physician, the process of rationalization carries with it a series of dehumanizing consequences. At or near the top of the list is the shift in control away from the physician and toward rationalized structures and institutions. In the past, private practitioners had a large degree of control over their work, with the major constraints being peer control as well as the needs and demands of patients. In rationalized medicine, external control increases and shifts to social structures and institutions. Not only is the physician more likely to be controlled by these structures and institutions, but he or she is also constrained by managers and bureaucrats who are not themselves physicians. The ability of physicians to control their own work lives is declining. As a result, many physicians are experiencing increased job dissatisfaction and alienation. Some are even turning toward unionization such as the Union of American Physicians and Dentists.[83]

From the patients' viewpoint, the rationalization of medicine causes a number of irrationalities. The drive for efficiency can make them feel like products on a medical assembly line. The effort to increase predictability will likely lead patients to lose personal relationships with physicians and other health professionals, because rules and regulations lead physicians to treat all patients in essentially the same way. This is also true in hospitals, where instead of seeing the same nurse regularly, a patient may see many different nurses. The result, of course, is that nurses never come to know their patients as individuals. Another dehumanizing development is the advent (at least in the United States) of "hospitalists," doctors who practice exclusively in hospitals. Now instead of seeing their personal physician (if they still have such a doctor), hospitalized patients are more likely to be seen by physicians whom they probably have never seen and with whom they have no personal relationship.[84]

As a result of the emphasis on calculability, the patient is more likely to feel like a number in the system rather than a person. Minimizing time and maximizing profits may lead to a decline in the quality of health care provided to patients. Like physicians, patients are apt to be controlled increasingly by large-scale structures and institutions, which will probably appear to them as distant, uncaring, and impenetrable. Finally, patients are increasingly likely to interact with technicians and impersonal technologies. In fact, because more and more technologies may be purchased at the drug store, patients can test themselves and thereby cut out human contact with both physicians and technicians.

The ultimate irrationality of this rationalization would be the unanticipated consequences of a decline in the quality of medical practice and a deterioration in the health of patients. Increasingly rational medical systems, with their focus on lowering costs and increasing profits, may reduce the quality of health care, especially for the poorest members of society. At least some people may become sicker, and perhaps even die, because of the rationalization of medicine. Health in general may even decline. These possibilities can be assessed only in the future as the health care system continues to rationalize. Since the health care system will continue to rationalize, health professionals and their patients may need to learn how to control rational structures and institutions to ameliorate their irrational consequences.

Dehumanized Death

Then there is the dehumanization of the very human process of death. People are increasingly likely to die (as they are likely to be born) impersonally, in the presence of total strangers: "A patient is every day less a human being and more a complicated challenge in intensive care to the consulting superspecialists . . . he is a case. . . . Doctors thirty years his junior call him by his first name. Better that, than to be called by the name of the disease or the number of the bed."[85]

This dehumanization is part of the process, according to Philippe Aries, by which the modern world has "banished death."[86] Here is the way Sherwin Nuland describes our need to rationalize death: "In recent generations, we have . . . created the *method of modern dying* . . . in modern hospitals, where it can be hidden, cleansed of its organic blight, and finally packaged for modern burial. We can now deny the power not only of death but of nature itself" (italics added).[87]

Similarly, Jean Baudrillard has written of "designer deaths," paralleling "designer births": "To streamline death at all costs, to varnish it, cryogenically freeze it, or condition it, put make-up on it, 'design' it, to pursue it with the same relentlessness as grime, sex, bacteriological or radioactive waste. The makeup of death . . . 'designed' according to the purest laws of . . . international marketing."[88]

Closely related to the growing power of physicians and hospitals over death, nonhuman technologies play an increasing role in the dying process. Technology has blurred the line between life and death by, for example, keeping people's hearts going even though their brains are dead. Medical personnel have also come to rely on technology to help them decide when it is acceptable to declare death. What could be more dehumanizing than dying alone amid machines rather than with loved ones?

When people are asked how they wish to die, most respond with something like this: quickly, painlessly, at home, surrounded by family and friends. Ask them how they expect to die, and the fear emerges: in the hospital, all alone, on a machine, in pain.[89]

Here is the way Nuland describes dehumanized death amid a sea of nonhuman technologies:

> The beeping and squealing monitors, the hissings of respirators and pistoned mattresses, the flashing multicolored electronic signals—the whole technological panoply is background for the tactics by which we are deprived of the tranquility we have every right to hope for, and separated from those few who would not let us die alone. By such means, biotechnology created to provide hope serves actually to take it away, and to leave our survivors bereft of the unshattered final memories that rightly belong to those who sit nearby as our days draw to a close.[90]

6

Dealing With McDonaldization

A Practical Guide

W hat can people do to deal with an increasingly McDonaldized world? The answer to that question depends, at least in part, on their attitudes toward McDonaldization. Many people view a McDonaldized world as a "velvet cage." To them, McDonaldization represents not a threat but nirvana. Weber's metaphor of an iron cage of rationalization communicates a sense of coldness, hardness, and great discomfort. But many people like, even crave, McDonaldization and welcome its proliferation. This is certainly a viable position and one especially likely to be adopted by those who have lived only in McDonaldized societies and who have been reared since the advent of the McDonaldized world. McDonaldized society, the only world they know, represents their standard of good taste and high quality. They can think of nothing better than a world uncluttered with too many choices and options. They like the predictability of many aspects of their lives. They relish an impersonal world in which they interact with human and nonhuman automatons. They seek to avoid, at least in the McDonaldized portions of their world, close human contact. Such people probably represent an increasingly large portion of the population.

For many other people, McDonaldization is a "rubber cage," the bars of which can be stretched to allow adequate means for escape. Such people dislike many aspects of McDonaldization but find others quite appealing. Like those who see themselves in a velvet cage, these people may like the efficiency, speed, predictability, and impersonality of McDonaldized systems and services. Such people may be busy and therefore will appreciate obtaining a meal (or some other McDonaldized service) efficiently. However, they also recognize the costs of McDonaldization and therefore seek to escape it when they can. Its efficiencies may even enhance their ability to escape from it. That is, getting a fast meal may allow them the time to luxuriate in other, nonrationalized activities.

These people are the types who, on weekends and vacations, go into the wilderness to camp the old-fashioned way; who go mountain climbing, spelunking, fishing, hunting (without elaborate equipment), antique hunting, and museum browsing; and who search out traditional restaurants, inns, and bed-and-breakfasts. Such people try to humanize their telephone answering machines with creative messages such as, "Sorry, ain't home, don't break my heart when you hear the tone."[1]

Although the bars may seem like rubber, they are still there, however. For example, a company that sells prerecorded, humorous messages now rationalizes the escape route for those who prefer creative answering machine messages. Thus, people can buy a machine with an impressionist imitating Humphrey Bogart: "Of all the answering machines in the world, you had to call this one."[2] Similarly, for many, home baking now includes the use of bread-baking machines, which do not produce a very good loaf but "do everything but butter the bread."[3]

The third type of person believes that the McDonaldized cage is made of iron. If the impregnability of the cage has not led such a person to surrender completely, he or she is likely to be deeply offended by the process but to see few, if any, ways out. Unlike the second type of person, these individuals see escape routes (if they see them at all) that provide only temporary respites, soon to fall under the sway of McDonaldization. They share the dark and pessimistic outlook of Max Weber—and myself—viewing the future as a "polar night of icy darkness and hardness."[4] These are the severest critics of McDonaldization and the ones who see less and less place for themselves in modern society.[5]

This chapter suggests actions open to each of these three types of people in a McDonaldized world. Those who think of McDonaldization as a velvet cage will continue to frequent fast-food restaurants and their clones within other sectors of society and will even actively seek to McDonaldize new, as yet unrationalized, venues. At the other extreme, those who think of it as an iron cage may want to work for the radical transformation of McDonaldized society, which might involve efforts to return to a pre-McDonaldized world or to create a new, non-McDonaldized world out of the rubble created by the fall of the golden arches. Primarily, however, this chapter is directed at those who think in terms of a rubber or iron cage and who are interested in ameliorating some of the problems associated with McDonaldization. In the main, the focus here is on more moderate ways of dealing with McDonaldization.

I begin by discussing some attempts at creating non-McDonaldized institutions. Second, I provide an overview of collective efforts to modify McDonaldized systems and limit their negative effects. Finally, I discuss a few examples of other, more individual ways people deal with a McDonaldized society.

Creating "Reasonable" Alternatives: Sometimes You Really Do Have to Break the Rules

The excesses of McDonaldization have led to the development of less rationalized alternatives. These alternatives do not put a premium on the efficient production of goods and services or the efficient processing of customers. They focus on high-quality products instead of large quantities. They revel in the unpredictabilities of their products and services. Instead of nonhuman technologies, they employ skilled human beings who practice their crafts relatively unconstrained by external controls. Hence, these are not McDonaldized settings for workers or customers.

Alternatives to rationalized settings exist in businesses and other social institutions. For example, food co-ops specializing in vegetarian and health foods offer an alternative to the supermarket.[6] The food is healthier than that in supermarkets (although some supermarket chains—e.g., Whole Foods—have sought to present themselves as alternatives, even though they and the vast majority of their products are highly McDonaldized), the shoppers are often members of the co-op and therefore actively involved in its management, and the employees are frequently more involved in and committed to their work.

In education, alternatives to highly rational state universities include small schools such as Hampshire College[7] of Amherst, Massachusetts, which at one time had the motto, "Where It's Okay to Go Outside the Lines." (The fast-food restaurants are not above using similar mottoes; for example, Burger King has used "Sometimes You Gotta Break the Rules," even though that's the last thing it wants people to do.) Hampshire College now touts its "personalized" education. According to its website, at Hampshire College, "Students design their own programs of study instead of following predetermined academic pathways (no 'off-the-shelf' majors here)."[8] There are no traditional academic disciplines at Hampshire, and students receive detailed written evaluations in narrative form rather than letter or numerical grades. Another example of a nonrationalized alternative to the McDonaldized enterprise is the bed-and-breakfast (B&B). In fact, one news report on B&Bs was titled, "B&Bs Offer Travelers Break From McBed, McBreakfast."[9] B&Bs are private homes that rent out rooms to travelers and offer them home-style hospitality and a homemade breakfast in the morning. Traditionally, the hosts live in the home while they operate it, taking a personal interest in the guests. Although B&Bs have existed for a long time, they began to boom at the beginning of the 1980s.[10] Some travelers had grown weary of the cold impersonality of rationalized motel rooms and instead sought out the types of nonrationalized accommodations offered by B&Bs. Said one visitor to a B&B, "It was marvelous. . . . The innkeepers treated us like family. It was so comfortable and friendly and charming and romantic."[11]

But success, once again, has brought with it signs of McDonaldization. The range of amenities offered at B&Bs is expanding, and the prices are rising. It is getting harder to distinguish B&Bs from inns or small hotels. Owners increasingly no longer live in the B&Bs but hire managers to run them. Said one observer, "Your best B&Bs are those where the owner is on the premises. . . . When the owner leaves and hires a manager, bad things start happening. Dust balls start accumulating under the beds, the coffee gets stale, and the toast is burnt."[12]

In other words, quality suffers. With the expansion of B&Bs, the American Bed and Breakfast Association came into existence in 1981, and guidebooks about B&Bs proliferated. Now inspections are being undertaken, standards developed, and a rating system implemented. In other words, efforts are under way to rationalize the burgeoning B&B industry.

The pressure to McDonaldize is even greater in England. The British Tourist Authority is pressing B&Bs that want an officially approved crown rating to offer a set of uniform amenities, such as full-length mirrors, ironing boards, telephones, televisions, and trouser presses. Such pressures led to the homogenization of B&Bs and to increasing difficulty in distinguishing them from motel and hotel accommodations. B&Bs are rated by how many amenities they have rather than more esthetic, subjective, and nonquantifiable factors, such as the warmth of the welcome, the friendliness of the atmosphere, the attractiveness of the setting, or the historical or architectural value of the building.[13]

As nonrationalized institutions become successful, pressures mount to McDonaldize them. The issue then becomes how to avoid rationalization. One thing to avoid is too much expansion. At some point, any institution will grow so large that it requires increasingly rational principles to function. With larger size comes another danger—franchising, which almost by definition brings with it rationalization. Because greater size and franchising hold out the almost overpowering lure of greater profits, the entrepreneurs behind a nonrationalized business such as a B&B must always keep the reasons for creating such a business in the forefront of their thinking. They must also keep in mind their obligations to the customers who frequent them because they are not McDonaldized. However, as creatures of a capitalist society, they might well succumb to greater profitability and allow their businesses to expand or be franchised. If they do, I hope they use their profits to begin new nonrationalized enterprises.

Fighting Back Collectively: Saving Hearts, Minds, Taste Buds, and the Piazza di Spagna

The preceding examples represent positive efforts to resist McDonaldization; however, more direct, often more negative, actions are also possible. If a number of people band together, they can form a movement against a specific component of the process (for example, McDonald's or Wal-Mart) or against the process in general.

One example of the success of collective movements is the elimination of the use of trans fats by most fast-food restaurants. While many of the ingredients in fast food have come under assault over the years, much attention came to be devoted to the fact that fast foods were usually high in trans fats (especially from partially hydrogenated oil). Trans fats are not only bad in themselves but can raise "bad" cholesterol (LDL) and lower "good" cholesterol (HDL), increasing the chances of having a heart attack. Trans fats are widely used in fast-food restaurants because they are cheap and can be stored easily for long periods of time, the oils that include them are very easy to work with, and they extend the shelf life of many products.[14] In other words, such oils are widely used because they are an ideal McDonaldized product. They are inexpensive (calculability), can be stored for long periods of time (calculability), are easy to work with (efficiency), and whatever is produced with them has a long shelf life (calculability).

While some countries (e.g., Denmark) passed laws limiting the amount of trans fat in food, the United States has relied on fast-food restaurants to take action. Local governments, however, have taken action, with New York City being the first to ban the use of trans fats in restaurants.[15] The state of California followed suit in 2010.[16] McDonald's was slower than many of its competitors (e.g., Wendy's) in reducing or eliminating trans fats.[17] However, by early 2007, even McDonald's had settled on a new trans fat–free formula for cooking its French fries,[18] and McDonald's is now 100% trans fat free.

The three most important examples of social movements against McDonaldization are the national and sometimes international campaigns against McDonald's (the McLibel Support Campaign), against fast food (Slow Food), and against Wal-Mart and other chains and superstores (Sprawl-Busters). Let us look briefly at each, as well as at a number of efforts by local communities to combat McDonaldization.

McLibel Support Group: McDonald's Pyrrhic Victory

The origins of the McLibel Support Campaign are traceable to a libel suit brought by McDonald's (United Kingdom) against two unemployed associates of London Greenpeace, Helen Steel and David Morris.[19] These individuals were involved in the distribution of a "fact sheet" that faulted McDonald's for many of the things that it stands accused of in this book (endangering people's health, damaging the environment, and offering poor working conditions and pay). The trial, which ended in January 1997, took more than 30 months to complete and became the longest running libel trial in England's history. The judge found for McDonald's on most counts but on several matters favored the position taken by the defendants. For example, the judge ruled that McDonald's exploits children, deceptively claims that its food is nutritious, and poses a risk to the health of its long-term customers. This victory was a Pyrrhic one for McDonald's, one widely seen as a public relations disaster. Later legal decisions (1999, 2005) supported the 1997 decision.

McDonald's spent around $15 million on the trial, hiring the best lawyers, whereas the impecunious Steel and Morris defended themselves. Of further embarrassment to McDonald's was that Steel and Morris continued to appeal the verdict.[20]

Several million copies of the original leaflet "What's Wrong With McDonald's: Everything They Didn't Want You to Know" have been distributed around the world, and it has been translated into a number of languages. More important, a site created on the World Wide Web (www.mcspotlight.org) reported at its peak an average of 800,000 "hits" a month.[21] McSpotlight has become the heart of a global movement in opposition to McDonald's as well as other aspects of McDonaldization.[22] It acts as the repository for information on actions taken against local McDonald's throughout the world and offers information on conducting such actions.[23] It is also the driving force in the annual Worldwide Anti-McDonald's Day.[24] Among "McSpotlight's" other targets is the Body Shop, accused of concealing behind its "green" image the fact that its products are detrimental to the environment, that it pays low wages, and that it encourages consumerism.

McLibel also supports the unionization of McDonald's workers, and there are occasional signs in the United States and elsewhere (e.g., China) of other efforts in this direction.[25] Although the initiative flies in the face of the general decline of labor unions, should it become successful, it would offer another base of opposition against McDonaldization. Fast-food restaurants have shown little inclination to deal with dehumanizing working conditions. Burger King, for example, has fought hard against unionization.[26] McDonald's is also well known for its hostility toward unions. For example, it destroyed a union in Moscow even after signing a collective agreement with it and shut down a store in Germany to avoid dealing with a workers' council.[27] (However, in early 2007, McDonald's agreed to the unionization of at least some of its restaurants in China.[28]) As long as a steady supply of people is willing to work in such settings for even just a few months, fast-food chains will not do much about their working conditions.

In some locales, McDonald's has faced an inadequate supply of workers from its traditional labor pool—teenagers. Rather than improve the work to attract more workers and keep them on the job longer, McDonald's has responded by broadening its hiring net. It now seeks out teenagers who live in distant communities, hires disabled adults, and brings in older employees, often retirees.[29] KinderCare also hires older people to make up for the shortage in younger workers. In fact, one expert said, "For old people who need to be needed, it [KinderCare] sure beats working in McDonald's."[30]

Slow Food: Creating a Place for Traditional, Regional, and High-Quality Food

The Slow Food movement had its origins in a mid-1980s grassroots movement organized by an Italian food critic[31] against the opening of a McDonald's in Rome (further discussed later in this chapter). It is opposed to the homogenization of food styles and states that its mission is "to give voice to local cooking styles and small-time food producers." It has also taken on the task of "fending off the homogenizing effects of European Union regulations on regional culinary treasures."[32] More positively, its objectives are "to provide members from all different countries with an identity."[33]

Slow Food (www.slowfood.com) has sought, quite successfully, to become a force throughout the world and now has more than 100,000 members in 150 countries.[34] It

- supports traditional ways of growing and raising food exceptional in quality and taste;
- favors the eating of such food as opposed to McDonaldized food;
- seeks to continue local traditions in how food is produced, what is eaten, and how it is prepared;
- favors food traditional preparation that is as close to handmade as possible;
- favors raw ingredients that are as specific to the place in which the food is made as possible;
- fights against environmental degradations that threaten local methods of producing food;
- supports the local shopkeeper and restaurateur (it favors "local inns and cafes");[35]
- creates local "convivia" throughout the world—there are more than 850 of them—that meet and engage in actions to further the above causes;
- has an "Ark of Taste" listing hundreds of foods that are endangered and in need of protection: "Ark foods must withstand the threats posed by bland, synthetic, mass-produced and menacingly cheap food";[36]
- seeks to involve restaurants, communities, cities, national governments, and intergovernmental agencies in the support of slow food;
- offers annual Slow Food awards, especially to those "who preserve biodiversity as it relates to food—people who may in the process save whole villages and ecosystems";[37]
- offers special prizes and support to Third World efforts and seeks to help organize local efforts to help conserve "prizewinners' plants, animals, and foods."[38]

In these and many other ways, Slow Food is fighting to sustain the continued existence of non-McDonaldized alternatives within the realm of food. Such alternatives in all realms and of all types need organizations like Slow Food and efforts such as these if they, and we, are not to be inundated further by a sea of McDonaldized phenomena. There is no reason why similar global organizations cannot be formed with the objective of sustaining non-McDonaldized alternatives and warding off the onslaught of McDonaldization in various realms.

While the maintenance and defense of non-McDonaldized alternatives are both important, it must be remembered that the Slow Food movement does not want to create a "museum." That is, it is not interested in simply maintaining the past and present but, rather, in creating the future. It is consequently important for it, and all organizations like it, to be actively involved in encouraging the creation of new, non-McDonaldized forms. This may involve making new combinations from what already exists or creating something entirely new. The latter is no easy task, but it must be kept in mind in the effort to maintain extant, non-McDonaldized forms.

This movement is clearly one of a very different order from the McLibel group. The impoverished targets of the McDonald's libel suit are a far cry from the mainly well-heeled gourmets drawn to Slow Food, its convivia around the world, and its regular meetings—Salone del Gusto—oriented to the appreciation of gourmet food. Most of the supporters of McLibel would be shocked by the prices for a meal in and around the Salone del Gusto. Slow Food focuses mainly on the issue of the poor quality of food (and, implicitly, almost all other products) in McDonaldized restaurants and food emporia, while McLibel focuses on threats to health, the environment, and the workers.

However, Slow Food does have a more populist side in which it seeks to support and reward small farmers, beekeepers, and other workers who continue to produce high-quality products in traditional ways. Slow Food is also very concerned about the environment. Whatever the differences in goals, methods, and social class of most of the participants, these groups share a hostility to the McDonaldization of society and are seeking to ameliorate its worst excesses, if not support and create alternatives to it.

In 2004, Slow Food sponsored the first Terra Madre meetings in Turin, Italy (it takes place every 2 years with the most recent in 2010). Nearly 5,000 delegates attended this first meeting (about the same number attended the 2010 meeting), representing 1,200 food communities from 130 countries. The key idea here is that of the "food community," which "consists of all those working in the food sector—from production of ingredients to promotion of end products—whose products are of excellent quality and produced sustainably. A food community is connected to a specific geographical area from a historical, social, economic and cultural point of view."[39]

There are two types of food communities. The first is *territory based*. In this case, although there may be diverse products, they are all derived from a given geographic area or native ethnic group. The second is *product based* and may include all farmers, breeders, processors, distributors, and others involved in producing a given product in a defined geographic area. Limited quantities are produced, and what is produced tastes good; does no harm to the environment, animals, or humans; and involves fair compensation and no discrimination or exploitation.[40] While Slow Food has always been concerned with production, it previously seemed to focus more on consumption than production. With Terra Madre, Slow Food signaled that it was concerned with *both* consumption and production, especially the production and consumption of less- or non-McDonaldized products.

Mention should also be made of the related Slow Cities movement, which seeks to bring many of these principles to bear on cities in Italy and beyond. This organization seeks to go beyond food and to protect art, architecture, a way of life—culture in general—from McDonaldization (and Americanization). The mayor of the Tuscan city of Greve stated, "The American urban model has invaded our cities and risks making Italian towns look the same. We want to stop this kind of globalization."[41] This mayor links the Slow Cities movement with sustaining alternatives to McDonaldization: "We can't stop large, fast-food chains [from] coming here if they request it, but we hope that people who come to our towns will not want to eat exactly the same hamburger they can eat in Melbourne, London or Paris, but want something genuine and different."[42]

The ideas behind the Slow Food Movement have spread to many other domains such as "slow travel and tourism."[43] Slow travel and tourism includes traveling less, more slowly, staying longer, and enjoying the travel experience itself rather than treating it merely as an efficient means to an end.

Sprawl-Busters: A "Hit List" of McDonaldized Superstores

Sprawl-Busters (www.sprawl-busters.com), founded by Al Norman, grew out of his successful effort to keep Wal-Mart out of his hometown of Greenfield, Massachusetts. The organization offers consulting services to local communities that want to keep out McDonaldized superstores and chains. For his efforts, the TV program *60 Minutes* called Norman "the guru of the anti-Wal-Mart movement."[44]

Among the services offered by Sprawl-Busters to local communities are help with overseeing media operations, raising money, petitioning for referendums, conducting data searches, and the like. Beyond Wal-Mart, organizations on Sprawl-Buster's "hit list" include Super Kmart, Home Depot, CVS, and Rite-Aid. The main objective is to keep such superstores and chains out to protect local businesses and the integrity of the local community. As of late 2011, they claim that at least 431 communities had prevented the entry of a big-box store.[45] Al Norman published *Slam-Dunking Wal-Mart: How You Can Stop Superstore Sprawl in Your Hometown.*[46]

Local Protests: Not Wanting to Say "Bye-Bye to the Neighborhood"

Some local communities have fought hard on their own, at times successfully, against the invasion of fast-food restaurants[47]—against the garish signs and structures, traffic, noise, and rowdy nature of some of the clientele drawn to fast-food restaurants. They have fought generally against the irrationalities and assaults on tradition that the fast-food restaurant represents. Thus, some of the communities highly attractive to fast-food chains (for example, Sanibel Island in Florida) have few, if any, fast-food restaurants.

The resort village of Saugatuck, Michigan, fought McDonald's attempt to take over the site of a quaint old café called Ida Red's. Said one local businessman, "People can see McDonald's anywhere—they don't come to Saugatuck for fast food." The owner of a local inn recognized that the town was really resisting the broader process of rationalization: "It's the Howard Johnson's, the McDonald's, the malls of the world that we're fighting against. . . . You can go to a mall and not know what state you're in. We're a relief from all that."[48]

Outside the United States, the resistance has often been even greater. The opening of the first McDonald's in Italy, for example, led to widespread protests involving several thousand people. The Italian McDonald's opened near the picturesque Piazza di Spagna in Rome adjacent to the headquarters of the famous fashion designer Valentino. One Roman politician claimed that McDonald's was "the principal cause of degradation of the ancient Roman streets."[49] Later, protests against the opening of a McDonald's in the medieval main market square of Krakow, Poland, led one critic to say, "The activities of this firm are symbolic of mass industrial civilization and a superficial cosmopolitan way of life. . . . Many historic events happened in this place, and McDonald's would be the beginning of the cultural degradation of this most precious urban area."[50]

The resort city of Hove is the largest town (91,900 inhabitants) in Great Britain without a McDonald's (although neighboring Brighton has four of them) or a Burger King.[51] Because of the resistance to the fast-food invasion, Hove has a thriving and diverse restaurant business, including, on its main street, "six Italian, five Indian and two French restaurants."[52]

Despite the passionate resistance to McDonaldized businesses in some locales, few communities have successfully kept franchises out completely. Similarly, small communities have usually been unsuccessful[53] in keeping out Wal-Mart stores, even though they usually devastate local businesses when they move in[54] and hurt the communities further on the rare occasions when they depart.[55]

In response to such protests and criticisms, and as an attempt to forestall them in the future, McDonald's is building outlets that fit better into the community in which they are placed. Thus, a McDonald's in Miami's Little Havana has a Spanish-style roof and feels more like a hacienda. Another in Freeport, Maine, looks like a quaint New England inn.[56] A McDonald's restaurant on Long Island stands in a restored, 1860s, white colonial house. The interior has a 1920s look.[57] The managing director of Polish operations said of a McDonald's in Poland, "We took a 14th-century building that was devastated and restored it to its natural beauty."[58] In Vienna, McDonald's opened McCafé, but it has been opposed by the Union of Vienna Coffee Houses, which fears the closing of the famed local coffee houses. The owner of one such café lamented, "You can never bring them back. . . . What we offer is an extension of the Viennese living room—a lifestyle."[59]

McDonald's has also shown signs of responding to environmental groups by creating less harmful packaging to the environment.[60] In 1990, McDonald's began eliminating its polystyrene "clamshell" hamburger box. The box had been attacked by environmentalists because its production generated pollutants. More important, the boxes lingered for decades in landfills or on the sides of roads. A paper wrapping with a cellophane-like outer layer took their place. In addition, McDonald's has purchased billions of dollars in recycled products and reduced packaging by an enormous amount. Following measures recommended by the Environmental Protection Agency's Green Lights program, they have started designing their buildings to be more energy efficient and have trained managers to reduce energy use in their restaurants. Furthermore, McDonald's corporation has entered into a partnership with Conservation International in an effort to focus on water and energy conservation and the protection and maintenance of animal and plant biodiversity.[61] Said one environmentalist, "I think that the public is pressuring these people into taking positive steps."[62] Today, McDonald's is pushing a broader environmental program called "Best of Green," which highlights its innovations in energy, packaging, anti-littering, sustainable food, and the like.[63]

Responses to complaints indicate that the fast-food restaurant is quite an adaptable institution, although all these adaptations remain within the broad confines of rationality. Some people have actually complained about the disappearance of the huge, old-fashioned golden arches, and at least one McDonald's franchise has responded by bringing them back. On the other hand, in response to complaints from upscale clientele about the dehumanizing dining environment, a McDonald's on 160 Broadway in Manhattan's financial district offers Chopin on a grand piano, chandeliers, marble walls, fresh flowers, a doorman, and hosts who show people to their tables. The golden arches at that restaurant are virtually invisible. However, besides a few classy additions to the menu (espresso, cappuccino, tarts), the fare is largely the same as in all other McDonald's (albeit with slightly higher prices). One visitor, underscoring the continuity between this franchise and all others, commented, "A smashing place, and the best thing is you can still eat with your fingers."[64]

Coping Individually: "Skunk Works," Blindfolded Children, and Fantasy Worlds

Individuals uncomfortable with or opposed to McDonaldization can challenge it in many ways. People who think of the bars of the rationalized cage as made of rubber may choose to extract the best of what the McDonaldized world has to offer without succumbing to its dangers and excesses. This is not easy to do, however, because the lure of McDonaldized institutions is great, and it is easy to become a devotee of, and enmeshed in, rationalized activities. Those who use rationalized systems for what they have to offer need to keep the dangers of McDonaldization always in the forefront of their thinking. But being able to get a bank balance in the middle of the night, avoid hospital emergency rooms by having minor problems cared for at "McDoctors," and lose weight quickly and safely at Jenny Craig, among many other conveniences, are all attractive possibilities for most people.

How can people take advantage of the best that the McDonaldized world has to offer without becoming imprisoned in that world? For one thing, they can use McDonaldized systems only when such use is unavoidable or when what they have to offer cannot be matched by nonrationalized systems. To help limit their use, perhaps we should put warning labels on the front doors of McDonaldized systems, much like those found on cigarette packs: **WARNING! Sociologists have found that habitual use of McDonaldized systems is hazardous to your physical and psychological well-being and to society as a whole.**

Above all, people should avoid the routine and systematic use of McDonaldized systems. To avoid the iron cage, they must seek out nonrationalized alternatives whenever possible. Searching for these niches is admittedly difficult and time-consuming. Using the various aspects of McDonaldized society is far easier than finding and using nonrationalized alternatives. Avoiding McDonaldization requires hard work and vigilance.

The most extreme step would be to pack up and leave the highly McDonaldized society of the United States. Many, if not most, other societies have long since embarked on the rationalization process or are about to, however. A move to another society might buy people some time, but McDonaldization would eventually have to be confronted, this time in a less familiar context.

Games, Knitting, and Nonrationalized Niches

Far less extreme than seeking to exit a McDonaldized society are efforts to cope by playing games, participating in knitting groups, and carving out nonrationalized niches in rationalized systems. One common finding in studies of workers is their use of games as a way of coping with the negative aspects of their jobs.[65] Workers in fast-food restaurants (and other McDonaldized work settings) are no different and, in fact, may be more prone to such games because they are usually quite young, often teenagers. One example involves fast-food workers throwing things (e.g., pickles) at one another. Another is competing to see who can make a cheeseburger faster. Then there is a version of a "chugging" contest, but instead of beer cans, the shake machine is used: "place your head under the spigot while a coworker pulled back the handle. A steady stream of shake more than a half-inch thick in diameter would then flow into the mouth, and either brain freeze or an overflowing mouth would make short work of even the most determined contestant . . . shake spilled out of the player's mouth and onto his face and clothes."[66]

Such games never deal with the fundamental dissatisfactions associated with work, in this case McJobs, but they do help ease the pain and pass the time at work a bit more joyfully. By the way, customers could emulate such workers and engage in similar games to make their visit to a McDonaldized system a bit more satisfying.

An interesting form of consumer rebellion against McDonaldization is the growing interest in knitting. In the rather widespread "Stitch 'n Bitch" groups throughout the United States, people get together not only to knit but also to interact with one another on a face-to-face basis. Said a leader of this movement, "For a lot of people, it's a way of rebelling against a larger consumer culture. You know where it's made, and you know what goes into making it."[67] Another commented, "One of the most political things you can do is to make something yourself."[68] In other words, by knitting something, you know that the end-product is *not* McDonaldized; hence, knitting can involve a political stance and action against that process.

The ability of workers to carve out a non-McDonaldized niche in a McDonaldized system tends to be related to their position in the occupational hierarchy. Those in higher-ranking occupations have a greater ability to create such niches than those in lower-status occupations. Physicians, lawyers, accountants, architects, and the like in private practice have the capacity to create such an environment for themselves. Within large organizations, the general (unwritten) rule for those at the top seems to be to impose rationality on others, especially those with little power, while keeping their own work as nonrational as possible. Rationalization is something to be imposed on others, especially on those with little power.

Some people in lower-ranking occupations are also in a position to be largely free of rationalization. For example, taxi drivers, because they work primarily on their own, are free(r) to construct a nonrationalized work life. They can generally go where they want, choose their passengers, and eat and take breaks when they wish. Similar possibilities exist for night guards and maintenance workers in automated factories. Those who work on their own, or in relative isolation within an organization, are also in a better position to create a non-rationalized work environment.

It is possible to find such nonrationalized work in some high-tech organizations that have created and encourage the use of "skunk works," where people can be insulated from routine organizational demands and do their work as they see fit.[69] Skunk works emphasize creativity and innovation, not conformity:

> They were creating almost radical *decentralization and autonomy,* with its attendant overlap, messiness around the edges, *lack of coordination, internal competition,* and somewhat *chaotic* conditions, in order to breed the entrepreneurial spirit. They had *forsworn* a measure of *tidiness* in order to achieve regular innovation. (italics added)[70]

The italicized terms in the preceding quotation would all be considered nonrational or irrational from the point of view of a McDonaldized society.

Nonrationalized times and places tend to be conducive to creativity. It is difficult to be creative in the face of incessant, externally imposed, and repetitive demands. Working in a nonrationalized setting serves not only the individual but many employers and society as well. All need a steady influx of creative new ideas and products, which are far less likely to emanate from rigidly controlled bureaucratic settings than they are from skunk works.

Even in highly rationalized organizations, people can carve out nonrationalized work spaces and times. For example, by finishing routine tasks quickly, a worker can leave himself or herself time to engage in nonrationalized, albeit work-related, activities. I am not suggesting that finding nonrationalized occupations or carving out nonrationalized spaces within McDonaldized organizations is easy. Nor am I suggesting that everyone can or should operate in a nonrationalized way all the time. But it is possible for some people, some of the time, to carve out nonrationalized niches for themselves.

I would not want to push this idea too far for several reasons:

- Rationalized organizations provide the resources and settings needed to do creative work and the outlets for that work. In other words, nonrationalized niches of creativity need the support of rationalized systems.
- No large-scale organization can exist if it is composed of nothing but such niches. The result would be organizational chaos.
- Not everyone wants to work, or is capable of working, in such nonrationalized niches; indeed, many people prefer their workdays to be highly routinized.

Thus, I am not arguing for a work world composed of nothing but creative occupational spaces. I do assert the need, however, for more nonrationalized niches in an otherwise highly rationalized world.

Another idea is for people to create and operate nonrationalized niche businesses in an otherwise highly McDonaldized environment. B&Bs are examples of such businesses. Local, independent restaurants have been able to find niches for themselves in an industry that is probably more McDonaldized than any other.[71] The owner of a small local chain (but a chain, nonetheless) of restaurants stated, "There are always going to be savvy restaurateurs who can beat the chains. . . . All is not lost."[72]

Starbucks, which already had 187 stores in New York City in late 2011, will open its 188th coffee shop in early 2012 on the site of a successful local coffee shop, the Bean. However, even though it has been evicted from its site, the Bean is not giving up. It is reopening across the street where it hopes to succeed by offering an alternative to Starbucks and its "hospitable monotony."[73]

A Range of Individual Actions: If All Else Fails, Save the Children

The following list contains suggestions for individuals who want to combat McDonaldization.[74] Some of these suggestions are offered "tongue-in-cheek," although the reader should not lose sight of the fact that McDonaldization is an extremely serious problem.

- For those of you who can afford it, avoid living in apartments or tract houses. Try to live in an atypical environment, preferably one you have built yourself or have had built for you. If you must live in an apartment or a tract house, humanize and individualize it.
- Avoid daily routine as much as possible. Try to do as many things as possible in a different way from one day to the next.
- Do as many things as you can for yourself. If you must use services, frequent nonrationalized, nonfranchised establishments. For example, lubricate your own car. If you are unwilling or unable to do so, have it done at your local, independent gasoline station. Do not, at all costs, frequent one of the franchised lube businesses.
- Instead of popping into H&R Block at income tax time, hire a local accountant, preferably one who works out of an office in his or her home.
- Similarly, the next time a minor medical or dental emergency leads you to think of a "McDoctor" or a "McDentist," resist the temptation and go instead to your neighborhood doctor or dentist, preferably one in solo practice.
- The next time you need a pair of glasses, use the local storefront optometrist rather than LensCrafters.
- Avoid Hair Cuttery, Supercuts, and other haircutting chains; go instead to a local barber or hairdresser.

- At least once a week, pass up lunch at McDonald's and frequent a local café or deli. For dinner, again at least once a week, stay at home, unplug the microwave, avoid the freezer, and cook a meal from scratch.
- To really shake up the clerk at the department store, use cash rather than your credit card.
- The next time a computer automatically dials your number, gently place the phone on the floor, thereby allowing the disembodied voice to drone on, occupying the line so that others will not be bothered by such calls for awhile.
- When dialing a business, always choose the option that permits you to speak to a real person.
- Seek out restaurants that use real china and metal utensils; avoid those that use materials such as Styrofoam that adversely affect the environment.
- Organize groups to protest abuses by McDonaldized systems. If you work in such a system, organize your coworkers to create more humanized working conditions.
- If you must frequent a fast-food restaurant, dine at one such as In-N-Out Burger.[75]
- If you are a regular at McDonald's, try to get to know the counter people (although that can be difficult given their high turnover). Also, do whatever else you can to humanize and subvert the system. For example, instead of hastening through their meals, many breakfast customers, especially among the elderly, form informal "breakfast clubs" and "come every day of the week to read their papers, chat, drink coffee, and gobble down an Egg McMuffin."[76] If breakfasts can be deMcDonaldized, why not other meals? Why not other aspects of the fast-food experience?
- Read the *New York Times* rather than *USA TODAY* once a week. Similarly, watch PBS news once a week, with its three long stories, rather than the network news shows with their numerous snippets or, heaven forbid, *HLN News* (formerly *CNN Headline News*).
- Avoid most finger foods. If you must eat finger foods, make them homemade sandwiches and fresh fruits and vegetables.
- On your next vacation, go to only one locale and get to know it and its inhabitants well.
- Never enter a domed stadium or one with artificial grass; make periodic pilgrimages to Fenway Park in Boston and Wrigley Field in Chicago.
- Avoid classes with short-answer tests graded by computer. If a computer-graded exam is unavoidable, make extraneous marks and curl the edges of the exam so that the computer cannot deal with it.
- Seek out small classes; get to know your professors.
- Don't go to any movies that have roman numerals after their names.

Regina Schrambling has developed a variety of strategies similar to those in the previous list for dealing with the health threats (especially *Salmonella*) posed by the rationalization of food production.[77] Schrambling recognizes that returning to the pre-rationalized method of raising chickens is not the answer. She argues that the "lifestyles" of such chickens, including "worm-grubbing," led to *Salmonella* problems even in the pre-rationalized days of chicken production. Nevertheless, she prefers to shop at farmers markets and to buy chickens raised the older way. She buys her eggs "in hand-packed boxes from the same New York State farmer." In her view, such eggs are fresher and cleaner than mass-produced eggs. She also purchases cantaloupes from farmers markets and refuses to buy them in supermarkets because they have been in transit so long that there is an increased risk of spoilage and disease (by late 2011, 18 people had died from *Listeria* linked to cantaloupes; more deaths were expected before the outbreak ran its course). Although rationalization has allowed people to eat fruits and vegetables year-round, it creates costs and dangers. As she puts it, such foods have been raised "in countries where we would never dare drink the water, where pesticides banned here are used freely." She thus buys fruits and vegetables only during their local seasons.

Schrambling argues that people need to understand the limited seasons for fruits and vegetables: "the strawberry crop is really as fleeting as fireflies, that sweet corn waits for no one; it's best when eaten within hours of leaving the stalk. There's nothing like the farmers' market in January, with only potatoes and squash and apples for sale, to give a deep new appreciation of nature's cycles. . . . [W]e can't have all of the food all of the time."[78]

Schrambling's position seems reasonable, even laudable, but the forces of McDonaldization continue to press forward. For example, science has recently discovered that genetically altered tomatoes can be prevented from producing the gas that causes them to ripen.[79] Thus, these tomatoes, and potentially many other fruits and vegetables, can be left on the vine until maturity instead of being picked early, can be shipped great distances without refrigeration and stored for weeks, and can then be ripened through exposure to ethylene gas when the retailer wishes to put them up for sale. If this technique proves viable commercially, people will, contrary to Schrambling, have many fruits and vegetables and cut flowers "all the time."

Similarly, the strawberry crop may now not be as "fleeting" as Schrambling maintains. The Driscoll strawberry, grown in Watsonville, California ("the strawberry capital of the world"), is big, glossy, and, most important, available year-round because of the favorable climate. Surprisingly, Driscoll strawberries "actually have some flavor, too."[80]

Given the attention that McDonaldized systems devote to marketing to children, it is particularly important that steps be taken to prevent children from becoming mindless supporters. Fast-food restaurants sponsor cartoon programs, have tie-ins with movies aimed at children, and offer numerous promotions involving toys. In fact, McDonald's has become "the world's biggest toymaker on a unit basis," commissioning about 1.5 billion toys per year, more than Hasbro or Mattel.[81] According to the president of a sales promotion agency, "In research, we've seen the kids are clearly motivated by the toy, not the meal."[82]

To protect children, try the following actions:

- Instead of using a "McChild" care center, leave your child with a responsible neighbor interested in earning some extra money.
- Keep your children away from television as much as possible and encourage them to participate in creative games. It is especially important that they not be exposed to the steady barrage of commercials from rationalized institutions, especially on Saturday morning cartoon shows.
- Lead efforts to keep McDonaldization out of the school system.
- If you can afford it, send your child to a small, non-McDonaldized educational institution.
- Above all, when possible, avoid taking your children to fast-food restaurants or their clones in other domains. If no alternatives are present (for example, you're on a highway and the only options are various fast-food chains), consider blindfolding your child until the ordeal is over. (Remember, some of these suggestions are only half serious.)

Freedom: If You Can't Cope, Can You Escape?

What if, as is likely, all your collective and individual efforts at coping with McDonaldization fail? What else can you do? Suicide is one possibility, but that does seem too extreme, even to me. However, some less radical but still effective ways of escaping rationalized society may exist.

One possibility is to flee to an area that has been designed as an escape area. In his book *Ways of Escape*, Chris Rojek discusses predesigned and prestructured escapes such as theme parks (for example, Disney World) and heritage sites (for example, Gettysburg battlefields), as well as less obvious and organized sites such as "black spots" (such as John F. Kennedy's grave site in Arlington Cemetery or Auschwitz) and "literary landscapes" (like Ernest Hemingway's haunts in Key West, Florida).[83] To the degree that such sites are not McDonaldized, they could serve our purposes as escape areas; however, many of them are being pressured to McDonaldize, especially when they are designed for, or attract, hordes of people. Disney World was McDonaldized from the beginning, and the others are growing more McDonaldized over time as they attract more people.

Rojek also analyzes tourist sites such as the beach and the wilderness as places that offer people "loopholes of freedom."[84] Again, to the degree that such sites remain largely or totally free of McDonaldization, they can serve as regions of escape. However, they, too, come under pressure to McDonaldize as soon as a significant number of people discover them as escape zones.

In *Escape Attempts: The Theory and Practice of Resistance to Everyday Life*, Stanley Cohen and Laurie Taylor outline a number of highly diverse ways to escape the routines of everyday life.[85] Although not all routines are the result of McDonaldization (sometimes people develop their own routines), an increasing number are part of this process. Many of the alternatives discussed by Cohen and Taylor are subject to McDonaldization, but two deserve some attention. The first is escaping within oneself, especially through fantasy. No matter how McDonaldized the setting, one can escape into an internal fantasy world of one's own making. Thus, while wandering through the highly McDonaldized fantasies created by Disney World, people can dwell on fantasies of their own creation, perhaps even inspired by the prefabricated fantasies around them. There is a tendency to buy into, and internalize, those prefabricated, McDonaldized fantasies, but they can be evaded within the depths of one's own fantasy world. Escape into one's own imagination does nothing about challenging or changing McDonaldized systems, but it does offer those who want it a viable way out.

Another possibility is what Cohen and Taylor call "trips to the edge" or what Michel Foucault calls "limit experiences."[86] These experiences are defined by "excess" and "outrage," often involving things like hard drugs. But drugs are not needed to act in an excessive and outrageous manner. One can, for example, "leave everything behind, travel light."[87] Here are some examples: Walk across the United States (avoiding highways, motel chains, and fast-food restaurants, among many other places), camp (using nothing that has labels like REI, L.L. Bean, Coleman, and Winnebago) on a mountain in Tibet, or take a year off and write that book or song or symphony you've always wanted to write (preferably using pad and pencil). Oh, you can use drugs or sex to excess if you like, but they are not the only ways to go to the limit or the edge not yet reached by McDonaldization. Just remember, although you might fall off (there may be no McDonaldized nets to save you), the journey should be exhilarating. But hurry, McDonaldization is *never* far behind.

Some Concluding Thoughts

A wide range of steps can be taken to cope with, or escape from, McDonaldization. I hold out little hope that such actions will reverse the trend toward McDonaldization, even if most of us were to employ them, but despite this fatalistic view, I think the struggle is worthwhile for several reasons. First, making the effort will mitigate the worst excesses of McDonaldized systems. Second, it will lead to the discovery, creation, and use of more niches where individuals and groups who are so inclined can escape McDonaldization for at least a part of their day or even a larger portion of their lives. Finally, and perhaps most important, the struggle itself is ennobling. In nonrationalized, individual, and collective struggles, people can express genuinely human reason in a world that, in nearly all other ways, has set up rationalized systems to deny people this expression.

Although I have emphasized the irresistibility of McDonaldization throughout this book, my fondest hope is that I am wrong. Indeed, a major motivation behind this book is to alert readers to the dangers of McDonaldization and to motivate them to act to stem its tide. I hope that people can resist McDonaldization and create instead a more reasonable, more human world.

Some years ago, McDonald's was sued by the famous French chef, Paul Bocuse, for using his picture on a poster without his permission. Enraged, Bocuse fumed, "How can I be seen promoting this tasteless, boneless food in which everything is soft." Nevertheless, Bocuse seemed to acknowledge the inevitability of McDonaldization: "There's a need for this kind of thing . . . and trying to get rid of it seems to me to be as futile as trying to get rid of the prostitutes in the Bois de Boulogne."[88] Lo and behold, 2 weeks later it was announced that the Paris police had cracked down on prostitution in the Bois de Boulogne. A police spokesperson stated, "There are none left." Just as Chef Bocuse was wrong about the prostitutes, perhaps I am wrong about the irresistibility of McDonaldization. Yet before I grow overly optimistic, it should be noted that "everyone knows that the prostitutes will be back as soon as the operation is over. In the spring, police predict, there will be even more than before."[89] Similarly, it remains likely that, no matter how intense the opposition, the future will bring with it more rather than less McDonaldization.

Even if McDonaldization grows more prevalent, I hope that you will follow some of the advice outlined in this chapter for protesting and mitigating its worst effects. Faced with Max Weber's iron cage and the image of a future dominated by the polar night of icy darkness and hardness, I hope that, if nothing else, you will consider the words of the poet Dylan Thomas: "Do not go gentle into that good night. . . . Rage, rage against the dying of the light."[90]

7

Globalization and the Possibility of the DeMcDonaldization of Society?

This book has focused on McDonaldization as an increasingly omnipresent process not only in the world of fast food, or even food more generally, but in many other aspects of the social world (family, religion, criminal justice, and so on).[1] That focus is retained in the first part of this final chapter where we look at the increasing global reach of the process of McDonaldization. We discuss both McDonaldization as a type of globalization, as well as an example of what I have called the "globalization of nothing" (see below).

The objective in the second part of this closing chapter is to take a critical look at the "grand narrative" of progressive McDonaldization, to examine whether there are important countertrends and even whether there are signs that we are moving in the other direction toward "deMcDonaldization." These issues will be dealt with in two subsections. First, we will focus on the rise of Starbucks and the issues of whether it is of such great importance that a distinct process of Starbuckization has emerged and threatens to replace McDonaldization and is thereby suggestive of deMcDonaldization. We will also turn to the Internet and offer a similar analysis of eBay and eBayization[2] and, most important, of the transition from Web 1.0 to Web 2.0 (of which eBay is a part) and analyze whether this realm offers the most definitive evidence of deMcDonaldization.

Globalization and McDonaldization

That McDonaldization is related to globalization is clear in the fact that McDonaldization, by definition, involves a process that not only is spreading throughout the United States but also encompasses more and more "of the rest of the world." In fact, the theory of McDonaldization articulated in this book has come to be seen by many as, at least in part, a theory of globalization. For example, O'Byrne and Hensby have recently offered an overview of eight approaches to the study of globalization, one of which is McDonaldization.[3]

The following is the definition of globalization employed in the ensuing discussion:

> Globalization is a transplanetary *process* or set of *processes* involving increasing *liquidity* and the growing multidirectional *flows* of people, objects, places and information as well as the *structures* they encounter and create that are *barriers* to, or *expedite,* those flows.[4]

The goal in the remainder of this section is to examine McDonaldization from the perspective of each element of this definition.

McDonaldization is transplanetary. As has already been discussed, McDonald's itself exists in well over 100 countries in the world, and the components of the process of McDonaldization have certainly made their way to many other countries.

McDonaldization is a process (or set of processes). The term *McDonaldization* itself suggests its processual nature. As made clear in the definition, it is a process that affects an increasing number of sectors of society, as well as of the world as a whole. More specifically, it is a process that brings with it increasing degrees of the basic characteristics of McDonaldization—efficiency, predictability, calculability, and control. While these are, in the main, desirable characteristics, they can have their negative side—the increasing irrationality of rationality.

McDonaldization is liquid. The metaphor of a liquid is used here to make the point that McDonaldization is not something that is solid and locked in place. Because of its liquidity, because it is a process, it can move more easily from institution to institution, from society to society.

McDonaldization involves multidirectional flows. Unlike other global processes (such as Americanization), globalization does *not* flow in one direction. While

McDonaldization clearly had its roots in the United States and continues to be centered there, other parts of the world have been McDonaldized, and their exports of McDonaldized systems (IKEA from Sweden, Pret A Manger from England, Pollo Campero from Guatemala) now flow to many other parts of the world, including the United States.

McDonaldization involves the global flow of people, objects, places, and information. McDonaldization in general not only flows increasingly easily in the global age, but so do its various aspects. In terms of the flow of people, managers of McDonaldized systems travel the globe visiting affiliates, but more important is the global flow of the consumers of what these systems have to offer. Thus, those who move about the world on business or as tourists are often looking for familiar settings; those that have the familiar characteristics of McDonaldized (restaurants or hotels that are part of global chains) systems are often very attractive to them.

Various objects flow throughout the world, especially those carried by the highly McDonaldized global package delivery companies such as Federal Express, UPS, and DHL. In addition, many of the things used in McDonaldized systems (e.g., the frozen beef in the hamburgers sold in global fast-food chains) travel great distances, sometimes halfway around the world.

Given their material character, it may be difficult to think of places as flowing around the world. However, in a sense they do. It is not the places themselves but the basic designs for them that flow globally. For example, there are many common elements in the structure of an IKEA, a McDonald's, or a Pret A Manger wherever in the world they may be found.

This relates to the main type of flow—information—associated with globalization and McDonaldization. In fact, the designs mentioned above are one type of information flow. Others would include e-mails and tweets among those involved in McDonaldized chains across the globe; the instantaneous flow of data on sales, income, stock on hand, and so on from distant locales to central offices; and marketing and advertising campaigns.

McDonaldized systems encounter various structures that serve as barriers to their flows as well as creating structures of their own that expedite flows. Globalization and McDonaldization are not simply about flows but also about structures that can either impede or expedite those flows. On one hand, the global spread of McDonaldization might be impeded by structures such as borders and tariffs, as well as by widely held and hardened beliefs that are critical of various elements of the process. On the other hand, McDonaldized systems create various structures—supply chains, warehouses, transportation systems—that expedite the flow of the process. In focusing on flows, we cannot ignore the importance of such structures.

In sum, then, McDonaldization fits well the definition of *globalization* employed here and is, in that sense, an aspect of the larger process of globalization. Thus, while I agree with O'Byrne and Hensby that McDonaldization is a form and a theory of globalization, I take this position for a very different reason. It is its fit with the definition of globalization, and not the fact that it brings with it uniformity and homogeneity, that makes McDonaldization a type of globalization. In fact, the very nature of the definition of globalization employed here militates against thinking of globalization in such simplistic terms. Globalization is highly complex with multiple, often conflicting outcomes because it is processual, involves so much fluidity, and encompasses so many different, often multidirectional, flows.

The Globalization of Nothing

Nothing and Something

The Globalization of Nothing puts McDonaldization in the broader context of globalization, as well as seeing McDonaldized systems as one type of "nothing."[5] The latter appears to be a shocking argument since McDonaldized systems are often considered to be quite "something," to be quite important, even special, by most people.

Nothing is a "social form that is generally centrally conceived, controlled and comparatively devoid of distinctive substantive content."[6] Any McDonaldized system, with the fast-food restaurant being a prime example, would be a form of nothing. However, many other examples (e.g., mass-produced products such as automobiles, flat-screen TVs, and iPhones) meet the definition of nothing that have little or no direct relationship to McDonaldization.

Let us look at the example of a chain of fast-food restaurants in terms of the basic components of the definition of nothing. First, as parts of chains, fast-food restaurants are, virtually by definition, *centrally conceived*. That is, those who created the chain and are associated with its central offices conceived of the chain originally and are continually involved in its reconceptualization. For their part, owners and managers of local chain restaurants do little or no conceptualizing on their own. Indeed, they have bought the rights to the franchise and continue to pay a percentage of their profits for it because they want those with the demonstrated knowledge and expertise to do the conceptualizing. This relative absence of independent conceptualization at the level of the local franchise is one of the reasons we can think of the chains of fast-food restaurants, and franchises in general, as nothing.

Just as those in the central office do the conceptualization for the local franchises, they also exert great *control*[7] over them. Indeed, conceptualizing and reconceptualizing the franchise yields a significant amount of control. However, control is exercised by the central office over franchises in more direct ways. For example, it may get a percentage of a local franchise's profits, and if its cut is reduced because local profits are down, the central office may put pressure on the local franchise to alter its procedures to increase profitability. The central office may also deploy inspectors to make periodic and unannounced visits to local franchises. Those franchises found not to be operating the way they are supposed to will come under pressure to bring their operations in line with company standards. Those that do not are likely to suffer adverse consequences, including the ultimate punishment of the loss of the franchise. Thus, local franchises can also be seen as nothing because they do not control their own destinies.

The third aspect of our definition of nothing is that it involves social forms largely *lacking in distinctive content*. The whole idea behind franchised fast-food restaurants is to turn out restaurants that are virtual clones of one another. They generally look much the same from the outside, they are structured similarly within, most of the same foods are served, workers act and interact in much the same way, and so on.

There is thus a near-perfect fit between the definition of nothing offered above and any set of McDonaldized systems. However, this is a rather extreme view since, in a sense, "nothing is nothing." In other words, all social forms (including fast-food restaurants) have at least some characteristics that deviate from the extreme form of nothing. That is, they involve *some* local conceptualization and control, and each one has at least *some* distinctive elements. To put this differently, all social forms have some elements of "somethingness."

Something is "a social form that is generally indigenously conceived, controlled, and comparatively rich in distinctive substantive content."[8] Neither nothing nor something exists independently of the other; each makes sense only when paired with, and contrasted to, the other.

If a fast-food restaurant is an example of nothing, then a meal cooked at home from scratch would be an example of something. The meal is conceived by the individual cook and not by a central office. Control rests in the hands of that cook. Finally, that which the cook prepares is rich in distinctive content and different from that prepared by other cooks, even those who prepare the same meals.

While nothing and something are presented as if they are a dichotomy,[9] we really need to think in terms of a continuum from something to nothing. Although a fast-food restaurant falls toward the nothing end of the continuum, every fast-food restaurant has at least some distinctive elements; each has some elements of "somethingness." Conversely, while every home-cooked meal is distinctive, it is likely to have at least some elements in common with other meals (for example, reliance on the same cookbook or recipe) and therefore have some degree of nothingness. No social form exists at the extreme nothing or something pole of the continuum; they *all* fall somewhere in between.

However, it remains the case that some lie closer to the nothing or something end of the continuum. Fast-food restaurants, and more generally all McDonaldized systems, fall toward the nothing end of the something-nothing continuum. If McDonaldized systems are clearly an example of nothing, how does that relate to globalization?

Glocalization and Grobalization

To many students of globalization, the central issue is the relationship between the highly interrelated topics of homogeneity-heterogeneity and the global-local. The now-famous concept of "glocalization" emphasizes the integration of the global and the local, as well as far more heterogeneity than homogeneity.[10–12] *Glocalization* is defined as the interpenetration of the global and the local, resulting in unique outcomes in different geographic areas. That is, global forces, often associated with a tendency toward homogenization, run headlong into the local in any given geographic location. Rather than one overwhelming the other, the global and the local interpenetrate, producing unique outcomes in each location.

This emphasis on glocalization has a variety of implications for thinking about globalization in general. First, it leads to the view that the world is highly varied and growing increasingly pluralistic. Glocal realities in one part of the world are likely to be quite different from such realities in other parts. This view leads one to downplay many of the fears associated with globalization in general (and McDonaldization more specifically), especially the fear of increasing homogeneity throughout the world.

Those who emphasize glocalization thus argue that individuals and local groups have great power to adapt, innovate, and maneuver within a glocalized world. Although they may be subject to globalizing processes, these powerful individuals and groups are not likely to be overwhelmed by, and subjugated to, them; they are able to glocalize them.

Whether the forces of globalization overwhelm the local is contingent on the specific relationship between the forces and counterforces in any given locale. Where the counterforces are weak, globalizing forces may successfully impose themselves, but where they are strong, a glocal form is likely to emerge that uniquely integrates the global and the local. Thus, to fully understand globalization, we must deal with the specific and contingent relationships that exist in any given locale.

The forces impelling globalization are not (totally) coercive. Rather, they provide material to be used in concert with the local in the creation of distinctive glocal realities. As an example, the global mass media (say, CNN or Al-Jazeera) are not seen as defining and controlling what people think and believe in a given locale but, rather, as providing additional inputs that are integrated with many other media inputs (especially those that are local) to create unique sets of ideas and viewpoints. McDonald's is seen as a glocal phenomenon because, for example, it adapts its menu to different locales and offers one or a few items (e.g., McSpaghetti in the Philippines) that are specific to a given area.

There is no question that glocalization is an important part of globalization, but it is far from the entire story. Although some degree of glocalization occurs with McDonaldization, another aspect of globalization relates far better to McDonaldization. That aspect of globalization is well described by the concept of *grobalization*,[13] or the imperialistic ambitions of nations, corporations, organizations, and the like and their desire, indeed their need, to impose themselves on various geographic areas.[14] Their main interest is in seeing their power, influence, and, in some cases, profits grow (hence the term *gro*balization) throughout the world. *McDonaldization is both a major example of, and a key driving force in, grobalization.*

Grobalization is largely antithetical to glocalization. Rather than emphasizing the great diversity among various glocalized locales, grobalization leads to the view that while there are differences within and between areas, the world is growing increasingly similar. This focus, of course, heightens the fears of those who are concerned about the increasing homogenization associated with globalization. In contrast to the view associated with glocalization, individuals and groups throughout the world are seen as having relatively little ability to adapt, innovate, and maneuver within a grobalized world. Grobalization theory thus sees larger structures and forces as imposing themselves on individuals, groups, and cultures throughout the world and a tendency to overwhelm their ability to create their own distinctive worlds.

In yet another stark contrast, grobalization involves social processes that are largely unidirectional and deterministic. That is, forces flow from the global to the local, and there is little or no possibility of the local having any significant impact on the global. As a result, the global is generally seen as largely determining what transpires at the local level; the impact of the global is not highly contingent on what transpires at the local level or on how the local reacts to the global. Grobalization thus tends to overpower the local. It also limits the ability of the local to act and react, let alone to act reflexively back on the grobal.

From the perspective of grobalization, then, global forces are seen as largely determining what individual(s) and groups think and do throughout the world. This view would accord far more power to grobal media powers such as CNN and Al-Jazeera or global corporations such as McDonald's and IKEA.

If we juxtapose the glocalization-grobalization and nothing-something continua implied in the preceding discussion, we come up with four major possibilities: glocalization of nothing, glocalization of something, grobalization of nothing, and grobalization of nothing. The discussion to follow will focus on the grobalization of nothing because it best describes McDonaldization as a global process.

The Grobalization of Nothing

A specific example of nothing is a meal at McDonald's. Throughout the world, there is little that is distinctive about any given McDonald's restaurant, the food served there, the people who work in these settings, and the "services" they offer. And, of course, there has been a very aggressive effort to grobalize McDonald's and its meals—to expand their presence throughout much of the world. The global expansion of McDonald's (and other fast-food chains) and its meals is therefore a near-perfect example of both nothing and grobalization, of the grobalization of nothing.

Within the context of globalization, there is a strong affinity between grobalization and nothing. Among the reasons for that affinity are the following:

1. There is a far greater demand throughout the world for nothing than something. This is the case because nothing tends to be[15] less expensive than something; more people can afford the former than the latter. McDonald's places great emphasis on its low prices, and this is true of most other McDonaldized businesses such as IKEA.

2. The comparative simplicity and lack of distinctiveness of nothing appeal to a wider range of tastes. The food at McDonald's is famous for its basic salty, sweet taste, as are the simple, clean lines of IKEA furniture.

3. Nothing is far less likely to bother or offend those in other cultures. Although it has aroused outrage in some cultures, McDonald's has shown the ability to fit into many different cultures.

4. Because of the far greater potential sales, much more money can be, and is, devoted to the advertising and marketing of nothing, thereby creating a still greater demand for it than for something. McDonald's spends huge sums on advertising and has been very successful at generating great demand for its fare. In contrast, a local restaurant offering food that is something in, say, Greece cannot afford to advertise and thus has little chance competing with McDonald's.

Given the great demand, it is far easier to mass produce and distribute the empty forms of nothing than the substantively rich forms of something (e.g., homemade Greek food, handcrafted pottery). Indeed, many forms of something lend themselves best to limited, if not one-of-a-kind, production. A skilled potter may produce a few dozen pieces of pottery and an artist a painting or two in perhaps a week, a month, or even a year(s). While these crafts and artworks may, over time, move from owner to owner in various parts of the world, this traffic barely registers in the total of all global trade and commerce. Of course, there are the rare masterpieces that bring millions of dollars, but in the main, these one-of-a-kind works are small-ticket items. In contrast, thousands, many millions, and sometimes billions of varieties of nothing are mass produced and sold throughout the globe. Thus, the global sale of fast food like Big Macs, Whoppers, and Kentucky Fried Chicken, as well as the myriad other forms of nothing, is a far greater factor in grobalization than the international sale of pieces of high art (for example, the art of Van Gogh) or of tickets to Lady Gaga's most recent world tour.

Furthermore, the economics of the marketplace demand that the massive amount of nothing that is produced be marketed and sold on a grobal basis. For one thing, the economies of scale mean that the more that is produced and sold, the lower the price. Almost inevitably, then, American producers of nothing (and they are, by far, the world leaders in this) must become dissatisfied with the American market, no matter how vast it is, and aggressively pursue a world market for their products. The greater the grobal market, the lower the price that can be charged, meaning, in turn, that even greater numbers of nothing can be sold globally.

Another economic factor stems from the stock market's demand that corporations that produce and sell nothing (indeed, all corporations) increase sales and profits from one year to the next. When corporations simply meet the previous year's profitability or experience a decline, they are likely to be punished in the stock market and to see their stock prices fall. To increase profits continually, the corporation is forced to search out new markets. One way of achieving that end is constantly to expand globally. Since something is less likely to be produced by large corporations, there is far less pressure to expand the market for it. In any case, given the limited number of these things that can be produced by artisans, skilled chefs, artists, and so on, there are profound limits on such expansion. This, in turn, brings us back to the pricing issue and relates to the price advantage that nothing ordinarily has over something. As a general rule, the various types of nothing cost far less than something. The result, obviously, is that nothing can be marketed globally far more aggressively than something.

Also, items that meet the definition of nothing generally can be easily and efficiently packaged and moved, often over vast areas. The frozen hamburgers and French fries that form the basis of McDonald's business are prime examples. Clearly, it would be much harder to package and move fresh hamburgers and freshly sliced potatoes, especially over large distances (that's one of the reasons that In-N-Out Burger has not moved beyond the western United States or become a global presence). The flat boxes that are used to ship the various pieces that go into the at-home construction of IKEA furniture are far easier and cheaper to ship than the crates that are used to ship expensive, handmade, and fully constructed furniture.

Furthermore, because the unit costs of such items are very low, it is of comparatively little consequence if they go awry, are lost, or are stolen. In contrast, such problems with a piece of handmade furniture or an antique vase are disastrous. It is far more expensive to insure something than nothing, and this difference is another reason for the cost advantage that nothing has over something.

In sum, the objective in this section has been to demonstrate that McDonaldization is part, and an example, of the broad process of globalization. More specifically, it is a prime example of one subtype of globalization—the grobalization of nothing. As in the rest of this book, this discussion points to McDonaldization as a continuing, if not increasingly powerful, force. While I think that is the case, the discussion in the remainder of this chapter shifts to signs of the weakening of McDonaldization and even to the possibility of the de-McDonaldization of society.

The DeMcDonaldization of Society

DeMcDonaldization involves a decline in the power and reach of McDonaldization. We discuss the issue of deMcDonaldization under two headings. First, we look at Starbucks as a powerful competitor to McDonald's and whether a process of "Starbuckization" can be seen as an alternative to, or even replacement for, McDonaldization. Second, and more important, we turn to the Internet, especially Web 2.0, to discuss whether the diversity in, and the great power of the user (user-generated content) on, the Internet indicates that this increasingly important domain is better seen as deMcDonaldized than McDonaldized.

Starbuckization

Is Starbuckization different enough from McDonaldization that it deserves its own distinctive label? If so, does this indicate that we are in the midst of a process of deMcDonaldization?

Starbucks' image is that it is "cutting edge."[16] This not only prioritizes Starbucks but also implies that McDonald's is no longer on top. If Starbucks is on the cutting edge of the fast-food industry, then perhaps it should be the model for the process discussed here. One customer explicitly acknowledged this new prioritization when she said, "Starbucks is the new McDonald's."[17] Among the indications of Starbucks' ascendancy are the following:

- Other businesses now owe their existence to Starbucks and seek to emulate it. Said the CEO of the nearly 500-store Caribou Coffee chain, "I got into the business because of what they [Starbucks] created."[18] In China, a small chain, Real Brewed Tea, aims to be "the Starbucks of tea."[19]
- Starbucks clones have emerged around the world. In Addis Ababa, Ethiopia, Kaldi's "has a Starbucks-like logo and Starbucks-like décor, and its workers wear Starbucks-like green aprons."[20]
- The *Economist* toyed with the idea of a "tall latte index" to play the same role as the "Big Mac Index" in comparing prices around the world.[21]
- A pastor said of the satellite locations created by megachurches, "It's kind of like going to Starbucks. You know the product you're going to get."[22]
- There is a book that delineates the "leadership principles" behind Starbucks' success and communicates them to aspiring businesspeople.[23]

What Has Starbucks Added to, or Removed From, the McDonald's Model?

Starbucks has moved away from the mediocre quality of McDonald's fare and toward the McDonaldization of higher-quality products, especially its coffee. Prior to Starbucks, the quality of American coffee was quite dismal. Starbucks has spearheaded a now-widespread trend (including at McDonald's) in the United States toward the appreciation and consumption of higher-quality coffee. It has done the same in other countries—most notably Great Britain (664 Starbucks)—where the quality of the coffee served before Starbucks' arrival was, if anything, even worse than in the United States. Starbucks has succeeded wildly in countries with a history of poor coffee and without a strong coffee culture (e.g., Japan, with more than 800 shops). However, it will be a major challenge for Starbucks to succeed as well (or at all) in societies that already have high-quality coffee and are famous for their rich coffee cultures (for example, Italy, France, or Turkey). As of mid-2009, there are still no Starbucks in Italy, but there are 46 of them in France, most in Paris, as well as more than 100 in Turkey.

In many ways, Starbucks' most important innovation has been to soften McDonald's hard-edged approach and image. For example, instead of the stiff and unwelcoming seats offered by McDonald's, Starbucks has its overstuffed armchairs and sofas. In fact, Howard Schultz explicitly seeks to distance Starbucks from McDonald's by associating it with Ray Oldenburg's more homey notion of a "third place": "We're in the business of human connection and humanity, creating communities in a third place between home and work."[24] Schultz associates several characteristics with Starbucks as a third place, including offering "a taste of romance," "an affordable luxury," "an oasis," and "casual interaction." Starbucks also encourages customers to come with their laptop computers and to connect to the Internet through a Wi-Fi connection (for a fee for most customers). They can also browse not only the newspapers for sale (or left behind by others) but also the books and CDs for sale, and they can stay as long as they wish. Regardless, even Schultz recognizes that Starbucks is "not yet the ideal Third Place," with its lack of adequate seating and the fact that even the relatively few customers who choose (or are able) to stay for a while rarely, in fact, get to know one another.[25]

In contrast, McDonald's focuses on the dangers of people staying too long. Indeed, the more general message from McDonald's has been that the customer is less than welcome. Customers are expected to eat their food fast and get out as quickly as possible. More recently, the message has become "don't enter the restaurant at all"! Instead, whenever possible, customers are expected to use the drive-through window, buy their food, and then leave immediately, taking their food and debris with them. *The ideal McDonald's customer is one who never sets foot in the restaurant.*

Starbucks has worked hard to create the opposite, or at least a very different, image—that customers are welcome, so welcome that they can stay for hours, if not all day. Of course, this message is more image than reality. While Wi-Fi is offered "from the comfort of your favorite cozy chair," it is costly.[26] As one journalist put it, "How welcome can one feel when staring at a meter that is running?"[27] Furthermore, if more than a handful of customers tried to take Starbucks up on its "invitation" to stay for hours, its shops would soon be inundated with people. There are a minuscule number of sofas, armchairs, tables, and seats for the enormous numbers of customers who stop at a Starbucks on a given day. Most customers pick up their coffee in the shop and leave immediately or use its drive-through window. In fact, 80% of Starbucks' customers are "to-go" customers.[28] Thus, almost all of Starbucks' customers behave in much the same way as many McDonald's customers. (In fact, the vast majority of McDonald's customers undoubtedly remain in the restaurant longer than Starbucks' customers.) John Simmons argues that "the vast majority of Starbucks customers made no conversation or social interaction in the store."[29] The difference between the two food chains is much more appearance than reality.

Thus, it could be argued that *Starbucks' major innovation has been in the realm of theatrics.* It is about "providing a great experience accompanied by coffee."[30] Starbucks has created a kind of stage set in which a few, unpaid performers (really paying customers) create the illusion of an old-fashioned coffee house. Although they may like watching the show, most customers do not want, or are unable because of time constraints, to be performers in it. Because of a lack of enough seating, they probably could not be those performers even if they wanted to be part of the show. They often do, however, feel good (assuming they don't use the drive-through) being the audience as they line up for their coffee and then leave with it. They may even imagine that some day they, too, might have the time to be part of the show and linger in one of those armchairs.

Starbucks has created great theater,[31] and in so doing, it has created a new way of doing something McDonald's pioneered—using customers as unpaid workers. McDonald's customers line up for their own food, deliver it to their tables, and then clean up after themselves. Starbucks' customers do all of these things and sometimes more—such as adding the desired amount of milk and sugar to their coffee. And a few of them do something that is in many ways more important for Starbucks—they stay to drink their coffee in the shop, thereby serving as unpaid performers in the ongoing show taking place at their local Starbucks. These "actors" include not only casual coffee drinkers but also people who use Starbucks as a place to meet dates or (with the Wi-Fi access) as a part-time office.[32] In fact, some people regularly schedule business meetings there. Starbucks encourages this sort of thing, but it can do so because only a small number of people are interested in working (or are able to work) there, and in any case, the possibility of working there is limited by the small number of tables and chairs. Even if they wanted to, the vast majority of customers (even businesspeople) could not stay.[33]

Another aspect of its theatrics involves the ways in which Starbucks presents itself as a kind, gentle, and caring corporation. They tell us[34] a great deal about how much they care about their 193,000 (as of 2010) employees. The company demonstrates this care through, for example, the provision of seemingly generous benefits (e.g., health care, pensions, stock options). Schultz claims that these actions help Starbucks maintain lower turnover than the rest of the industry.[35] This practice stands in stark contrast to McDonald's, which has often been criticized for employing young, part-time workers and for providing them with few, if any, benefits. Then there is their very public concern for the coffee growers, as expressed in their interest in fair-trade coffee. Again, the contrast is with McDonald's, which is often accused of exploiting the land and those who provide them with beef, chicken, potatoes, and so forth.

There is growing evidence, however, that this contention, too, is more myth than reality; it is more public theater, or perhaps better public relations. In the examples mentioned above, Starbucks does *not* treat its employees as well as it would like us to think, and its commitment to fair-trade coffee is minimal. According to one source, less than 4% of Starbucks' coffee is fair-trade coffee, it rarely offers fair-trade varieties as its coffee of the day, and it has not followed through on, or publicized widely, its promise to brew such coffee on demand for customers who request it. Starbucks' commitment to the small coffee farmer is much more myth than reality, and other chains (Costa and Caffe Nero in Great Britain) do much more to ensure that fair-trade coffee is sold.

Should the Concept of "Starbuckization" Replace "McDonaldization"?

Are the successes of Starbucks, and the innovations associated with it, enough to conclude that "Starbuckization" should replace "McDonaldization" as *the* term to describe the various changes discussed in this volume? I think the answer is a clear and resounding no! After all, McDonald's was the "pioneer" in the fast-food industry and in bringing the process of rationalization to its customers and to the process of consumption. Since we reject the replacement of McDonaldization with Starbuckization, we therefore reject the idea that the success of Starbucks is an indication of deMcDonaldization.

Starbucks clearly fits, and operates in accordance with, the principles of McDonaldization. Starbucks is:

- *Efficient.* Customers line up in the shops for their coffee (and perhaps some food), pay, make their way to the cream and sugar stand, fix their coffee to taste, and in most cases dutifully leave; the drive-throughs at those Starbucks that have them offer the same kind of efficiency that is provided by the McDonald's drive-throughs. Since brewing and serving coffee are, by far, the major activities at Starbucks, they can be done far more efficiently than in McDonald's where a couple of dozen different food and drink products are "made." And, of course, the instant coffee, VIA, allows for the more efficient preparation of a cup of coffee at home or in the office.
- *Predictable.*[36] The logo, the shops,[37] the counters, the marquees listing available drinks, the coffee (especially the more exotic versions), the green aprons, the methods of preparation, the stand with cream and sugar, the carafes (with half-and-half, whole milk, and skim milk), the array of other products, the shelves with goods for sale such as coffeemakers, mugs, books, CDs, and so on are all pretty much the same from one locale to another.
- *Calculable.* While there is no pretense of low cost, the emphasis on quantity is clear in the coffee sizes, especially the large sizes. There is *no* small Starbucks coffee; the smallest—in the best tradition of "newspeak" in George Orwell's *1984*— is a "large!" (The other alternatives, using faux Italian names—and more theatrics—are venti and grande; the latter certainly communicates an even larger-sized cup of coffee than the tall—it is made to seem taller than tall.) Also indicative of calculability are the number of "shots" of espresso one can get as well as the (high) prices associated with the various drinks and sizes. In fact, while McDonald's associates itself with low prices, Starbucks has succeeded in getting people to pay very high prices, sometimes $4 (or more)—as a result, some call Starbucks "Fourbucks"[38]—for some exotic varieties of its coffee. In fact, the high prices add to Starbucks' effort to position itself at the high end of the market. And there is theater involved here, too. This is a high-end, "classy" show, not the cheap and garish one on view at the McDonald's down the road.
- *Controlled by nonhuman technology.* As in McDonald's, the system at Starbucks exerts great control through nonhuman technologies. For example, creative baristas

have been replaced by (mostly) automatic espresso makers (with a corresponding loss of skill and magic). Although Starbucks' theater may lead customers to want to stay, lingering, as we have seen, is impossible for most.

The Irrationality of Rationality at Starbucks

Starbucks has most, if not all, of the irrationalities associated with McDonald's (homogenization, disenchantment, dehumanization), as well as additional ones, such as the much higher prices than at McDonald's.

Customers are paying an extraordinary amount of money for what is essentially flavored water; the profit margins on each cup of coffee are enormous, and those margins are even higher in the case of the high-priced and more exotic drinks. There is a huge markup on a cup of coffee—perhaps as high as 95%.[39] Thus, a venti coffee that costs the consumer more than $2.00 might cost Starbucks as little as 10 cents to make. While Karl Marx wrote about the exploitation of the worker in the era of the domination of production capitalism, the focus has now shifted (although the worker is still exploited, as well) to the exploitation of consumers through not only unpaid labor but also extraordinary markups and price inflation.

The great expansion of Starbucks means that it must buy huge quantities of coffee beans, and that flow must continue uninterrupted. One negative associated with this need is that it cannot necessarily always get the best beans; indeed, the best beans may be grown in such small quantities that they are of little interest to Starbucks, given its enormous requirements. Despite its claim to be selling high-quality coffee, many experts complain about its mediocrity.[40]

While Starbucks has eschewed franchising, it has increasingly embraced joint ventures and licensing. These contradict Schultz's contention that franchising is a "forbidden word" at Starbucks.[41] The number of licensed Starbucks has increased rapidly (Safeway, Barnes & Noble, etc.), and now over a third of Starbucks in the United States are licensed rather than being run by Starbucks. And overseas there are more licensed and joint-venture Starbucks than those that are company owned. It may be that Starbucks has *less* control over licensees than they might have had over franchisees.[42] The quality of what Starbucks has to offer has been compromised by licensing.

The problem of homogenization is reflected in a statement by Howard Schultz about the burgeoning number of Starbucks in China (over 350 in 2011 but likely to grow dramatically in a short time): "Our stores there are a mirror image of what you see in . . . Pikes Place."[43] In Taiwan, the fear is that Starbucks is playing a major role in the disappearance of its distinctive tea shops and tea culture.[44]

There is also considerable uniformity to the coffee shops themselves both in the United States and the rest of the world, although in recent years,

Starbucks, like McDonald's, has grown increasingly sensitive to this issue, and there is more variability in store designs and greater efforts to adapt to local environments.

Starbucks often drives small, local, and unique coffee shops out of business and, in fact, has been sued for antitrust violations.[45] (However, the opposing view is that Starbucks has helped small coffee shops by raising awareness of higher-quality coffee, helping to create an interest in coffee culture, pioneering high prices for coffee, and providing niches where smaller shops can survive within the context of the expansion of Starbucks.[46])

An issue at Starbucks is the nutritional value of many of its products. This is particularly the case with creamy, sugary, Frappuccino-type drinks.[47] Such drinks have about the same (large) number of calories and fat as a McDonald's milkshake.

Although it has a public policy of not marketing to children and of not hooking children on addictive coffee drinks, its "starter drinks"—such as a banana mocha Frappuccino and caramel macchiato—inevitably attract pre- or early teens because they have similarities to milkshakes and Slurpees. The types of music promoted also attract children and teenagers to Starbucks.[48]

This is another way in which Starbucks has stood on the shoulders of McDonald's. It could be argued that McDonald's "infantilized" food.[49] It created a circus-like atmosphere and used a clown (Ronald McDonald) and other cartoon-like characters ("Hamburglar") to lure children who, in most cases, brought their parents along. It offered "finger food" that could be eaten with one's hands. It also offered very basic tastes—most generally the very sweet and highly salty, sometimes in the same food. Then there were the vast numbers of toys given away or sold, as well as the tie-ins to movies such as those made for children by Disney.

But how does this apply to Starbucks, which is oriented to adults and sells a largely adult-oriented range of coffees? The appeal to young people and young adults of its higher-priced and exotic coffees is the source of much of its success and, undoubtedly, its greatest profits. Many of the drinks are very sweet, highly flavored, very creamy, and/or quite frothy. They are, in the main, *coffee for those who don't really like coffee*. Starbucks has infantilized coffee in order to broaden the market for its products.

In sum, Starbucks has been a phenomenal success, but that success has been built on the model pioneered by McDonald's. While it has made a number of innovations, it is better seen as another example of McDonaldization than as a new model that has earned an appellation of its own. The most important conclusion to be derived from this discussion is that Starbuckization is by no means an indication of deMcDonaldization.

The Internet and DeMcDonaldization

Two interrelated aspects of the Internet will be discussed in this section. First, we will discuss whether eBayization is replacing McDonaldization and thereby is indicative of deMcDonaldization. Second, we will look at the transition from Web 1.0 to Web 2.0 (which subsumes eBay) and discuss whether a process of deMcDonaldization is involved in this transformation of the Internet.

eBayization

Elif Izberk-Bilgin and Aaron Ahuvia[50] argue that, while McDonaldization might have been an apt term to capture the reality of a material, Fordist world of consumption that existed a quarter of century ago, it does not well describe contemporary realities. Among the changes are post-Fordist production techniques, more fragmented consumers and consumer tastes, and a growing desire for more individualized products and services. eBayization is proposed as a paradigm for this new world, especially as it relates to consumption. eBayization has three basic dimensions: *variety* (as opposed to the limited offerings at McDonaldized fast-food restaurants), *unpredictability* (as opposed to the predictability of McDonaldized systems), and *market-mediated control* (rather than the centralized control of a McDonaldized system like McDonald's). eBayization is a very attractive idea, at least as it applies to the Internet and consumption. Since it is likely that consumption (and much else) will occur increasingly on the Internet, there is a strong case to be made for eBayization as the paradigmatic process of our time.

There is no question that the three characteristics associated with eBayization not only differ from those that define McDonaldization but seem to point to less rather than more McDonaldization. McDonaldized systems were, and are, designed to limit or even eliminate variety and unpredictability and to seek to better control the market. However, the basic characteristics of eBayization are superficial, and when probing beneath the surface, one finds the characteristics of McDonaldization that underpin it. More strongly, it is the McDonaldization of the basic structure of eBay that makes variety, (seeming) unpredictability, and (apparent) market-mediated control possible.

The variety of products offered on eBay "can boggle even a jaded mind," but Izberk-Bilgin and Ahuvia neglect to analyze the structure of eBay that makes this great variety possible. It may be difficult to visualize a McDonaldized structure underlying the variety of items listed on eBay. However, that structure is the database itself and its interface with buyers and sellers on the site. Variety attracts buyers and sellers alike, but that variety would not be manageable without the McDonaldized architecture of the database and the website that brings these items together in one efficiently searchable space. Without a

McDonaldized structure to classify and to access the large number and wide range of entries, as well as to make entering and ordering easy, eBay's variety would quickly degenerate into chaos.

There *is* unpredictability on eBay, but it, like variety, is made possible by the system's highly McDonaldized structure. The highly unusual and unique products found on eBay could not be sold and purchased, at least in large numbers to a wide range of consumers, were it not for its McDonaldized structure. eBay gives no more space to one item than the next and does not promote one over the other. Each item on its database is searchable under a preset number of keywords and appears within the same layout customizable only in limited and unvarying ways. Thus, the elements of unpredictability are made tolerable by the frame and structure of predictability that surround it.

Furthermore, it could be argued that these limited elements of unpredictability serve to "enchant" what is otherwise a highly disenchanted system. McDonaldized systems often conceal their rationalized core and make it more acceptable by surrounding it with enchantment and magic.[51] The seeming unpredictability on eBay performs a similar function. Consumers are drawn to both its unpredictable offerings and the unpredictability of obtaining things at bargain prices. This seeming magic serves to obscure the McDonaldization at the heart of eBay as well as to make it more tolerable.

Market-mediated control is an alternative to the strategy of centralized control employed by McDonald's.[52] However, its control calls into question the meaning of market mediation on eBay. In fact, of course, the whole idea of a market is that it be "free." However, the structure of eBay limits the free market in various ways. It is one user's responsibility to decide whether to undertake the risk of buying or selling with any other user, but eBay seeks to allay the fears of those who might be deterred by the risks associated with such transactions. This is done through the eBay user rating system, which gives buyers a number by which they can judge the reliability of a seller and the potential risks associated with a given transaction. Then there are the standardized procedures for filing complaints against other users so that buyers and sellers are provided with another means (number of such grievances) of evaluating each other's credibility. These sorts of things constitute interference in the free market of eBay, but they offer users a considerable amount of predictability in what appears on the surface to be an unpredictable world.

eBayizing McDonaldization?

eBay's chief goal is profit, and to maximize profit, it must be efficient. While the goal for any capitalistic enterprise is profit, the means of achieving maximum profitability may appear on the surface quite different according to the context of the enterprise's operations. eBay's means to the end of profitability requires listing as many entries on its database as possible, thereby making more likely the sale of many items. For eBay to make the sale of entries in its database efficient, it must appeal to as many potential entry-makers and searchers as possible, get them to register relevant information with the site, try to make sure they spend large amounts of time interacting with the database, and make converting an entry into a sale easy and low in risk. This ease of the transition from entry to sale also encourages users to enter items for sale and to buy items listed for sale by others.

While eBay is unpredictable due to the surprise factor found in the items for sale and their prices, the processes that surround the entry and sale of each item are highly predictable. Entirely predictable is the appearance of each entry as it is listed for sale. What is listed in the "box" might be anything, but the box itself is always the same. Similarly predictable are the filters that appear when a user searches within a particular category—for instance, when searching clothes, filters may be set to sort for size, color, style, brand, or condition.

Further increasing predictability is the fact that the seller of an item listed toward the top of any search result will be a "Power-Seller." To be included in that category, one has to have paid a fee after having been involved in at least 100 transactions that produced at least $3,000 in sales during the preceding 12 months. Furthermore, over the course of that period, the seller had to have received a rating of 98%. The Power Seller Program promises that dedicated and predictable eBay sellers will have increased visibility in search results, including top-of-the-list placement in searches. It is increasingly less likely that sellers fit the image of an unpredictable weekend yard-sale host.

In terms of calculability, the eBay model allows buyers to access an extraordinarily large number of items. Furthermore, each seller pays an insertion fee for the listing of an item, as well as a final value fee once the item is sold. The eBay site even hosts a value calculating tool, yielding results on the average price, full price range, and number of listings sold within a particular category during the preceding 3 weeks. Items, descriptors, and users are all coded by their own algorithms, enabling the connection of search to item. These complex equations allow items, terms, and users to be identified from the database instantaneously. Sellers are therefore encouraged to create effective listings by buying additional descriptors beyond the limited number allowed in a standard listing. Sellers may also block particular users from viewing a listing by entering specific user names, who will then be identified from the database through their particular algorithm. More generally, the focus of eBay is on maximizing the quantity of items listed and sold, as well as of the number of users.

We can find other websites (e.g., www.freecycle.com) with a great variety of items for sale, trade, or giveaway—websites that are not McDonaldized to a high degree. However, should such sites wish to strengthen and extend their reach, they will need to McDonaldize their underlying structure much the way that McDonald's—and eBay—have rationalized theirs. This is an important correction to the eBayization hypothesis: User-generated variety not only coexists with but thrives because of the rationalized structure of eBay. In sum, eBay, at least in its underlying structure, *is* McDonaldized and therefore should not be seen as an indicator of deMcDonaldization.

Web 1.0 and 2.0*

Web 1.0 can be described as the Internet in its first decade (the 1990s). That which exists as of this writing can, in the main, be seen as Web 2.0. However, Web 1.0 and 2.0 are viewed here mainly as temporally overlapping phenomena. *It is the explosion of user-generated content that defines Web 2.0 and differentiates it from the provider-generated content of Web 1.0.*

Examples of Web 1.0 include the Apple Store and other shopping sites that dictate the content and users' browsing and Fodors.com, which uses its own tastemakers to point tourists to various hotels, restaurants, and so forth. In contrast, Web 2.0 accords far less power to the creators of these systems and much more to their users; Web 2.0 sites, or at least the material on them, are, to a large extent, user generated. It permits the greatly increased ability to network with others socially. One way of describing this is to argue that the implosion of the consumer and the producer on Web 2.0 has led to the preeminence of the prosumer. That is, on Web 2.0, users produce that which they consume (e.g., users both produce and consume the profiles and networks on Facebook).

Major examples of Web 2.0, and of the centrality of the user-generation on them, include Wikipedia, where users generate and edit articles; Facebook, MySpace, and other social networking websites, where users create profiles to interact with one another and build communities; Second Life, where users create the entire virtual environment; the blogosphere, where participants read and write blogs; eBay and Craigslist, where buyers and sellers create the market; YouTube, Flickr, and other media sharing sites, where mostly amateurs upload and download videos and photographs; and Linux, a free, collaboratively built, open-source operating system, and other open-source software applications, like Mozilla Firefox, that are created and maintained by those who use them.

This explosion of user-generated content has dramatically transformed the Internet. There are many different ways to describe the nature of this change, to describe what is new and unique about Web 2.0, including the populist notion that many minds are better than one (the "wisdom of the crowds"), the

---

*The section was co-authored by Nathan Jurgenson and is based, in part, on George Ritzer and Nathan Jurgenson. "Production, Consumption, Prosumption: The Nature of Capitalism in the Age of the Digital 'Prosumer.'" *Journal of Consumer Culture* 10, no. 1 (2010): 13–36.

view that emphasizes the productivity and originality of mass self-expression, or the cyber-libertarian notion of the advantages that accrue from breaking down barriers and structures online (creating a "flattened world").

Like material realities such as fast-food restaurants, Web 1.0 was, and is, highly McDonaldized. Web 1.0 websites are constructed in a one-size-fits-all model, and from the point of view of those who own, control, or work on them, this makes the sites highly efficient to create, maintain, and use.

User efficiency is enhanced by the predictability of websites on Web 1.0; they are more or less identical from one time or place to another because content follows a predictable top-down pattern. The ubiquity of Web 1.0 sites like Yahoo! or the services of AOL serve(d) to eliminate inefficiencies associated with having to deal with different or changing websites. Because user choices are limited or nonexistent on Web 1.0, there is also great predictability for those in control of the websites. The major source of unpredictability in any McDonaldized system—human behavior—is largely eliminated from these websites, especially those associated with Web 2.0. In addition, once a website is created, it can remain in place indefinitely, further enhancing predictability from the perspective of the website's creators and users.

Calculability is no problem on Web 1.0 since those in control of the sites can *easily* monitor their use and calculate precisely variables such as the number of users and how the sites are used. Similarly, users can compare available sites to assess which ones allow them to use their time on them most efficiently.

Web 1.0 is, of course, controlled almost completely by the websites that are, in effect, nonhuman technologies. Once in place, these websites control what users do on them and give users few options. Since the websites are largely static, and the goal is to keep them that way, control is exercised over those who produce content on the sites.

Thus, a strong argument can be made that Web 1.0 is McDonaldized to a high degree. However, anything that is McDonaldized to such a degree is subject to the irrationality of rationality. One such irrationality that stands out in this case is dehumanization. On one hand, the humans who work on, or for, Web 1.0 websites are highly limited in what they can do; they cannot fully exploit their creative human capacities to improve the sites or respond as fully as possible to user needs and complaints. In addition, these websites are largely dehumanized from the point of view of their users. If they want to use a site, people must use it in the way its designers and operators intend. They cannot use their skills and abilities to alter the site or to use it in highly creative ways. Furthermore, the sites are structured in ways that are relatively uncollaborative. This is irrational, an irrationality of rationality, in the sense that Web 1.0 squanders its ability to make use of the skills and abilities of both those who work for the sites and those who use them. It is especially in the latter case that Web 2.0 has a huge advantage over Web 1.0. In one sense, Web

2.0 has reduced or eliminated the irrationalities of rationality associated with Web 1.0. In another sense, it could be argued that it has greatly heightened the rationality of these systems by figuring out how to get the most out of the people who use the sites without allowing them to compromise the basic functioning of the system. In this way, while Web 2.0 can be viewed as a rational next step, often pushed by profit-based motives, it exists partially outside the principles outlined by the McDonaldization thesis, and thus Web 2.0 can be seen as having, to some degree, a tendency toward *de*McDonaldization.

*De*McDonaldizing the Web? In many ways Web 2.0 is *less* efficient than Web 1.0, especially for users. The amount of time and energy users spend producing content on social networking sites, as well as blogging and microblogging (e.g., Twitter), writing comments on others' blogs, writing reviews on sites like Amazon, scanning items on offer on eBay, and so on far exceeds the amount of user-generation that existed on Web 1.0. Because it demands a massive amount of user-generated input, Web 2.0 tends not to be as efficient as Web 1.0, where content is generated by those associated with the site. Since Web 2.0 is defined by the ability of the masses to create content online, the general abundance of profiles, reviews, comments, opinions, news, photos, videos, and much else seems wasteful from the perspective of an efficient system. How many users ultimately contribute, or how much time they spend on their contributions, matters little. Instead, the focus is on the quality of what they produce (leaving aside the debate on the actual quality of Web 2.0 content such as Wikipedia entries). That Web 2.0 involves a focus on output irrespective of the amount of input is an example of its relative *in*efficiency.[53]

All of this also means that there is far more *un*predictability on Web 2.0 sites than on Web 1.0 sites. As is the case with eBayization, the basic structure of many sites on Web 2.0 is predictable (e.g., the nature of a Facebook page), but what does or does not find its way onto that page is largely unregulated and unpredictable. There are limits that vary by site, but they are quite wide, with the result that users are unable to predict what they will find when they log on to most Web 2.0 sites.

It is also much harder to quantify, to calculate, exactly what is transpiring on a Web 2.0 site. In part, this is because there is so much more going on, and it takes so many different forms. More important, while Web 1.0 sites tend to be restricted to objective matters (did one order something? how much was paid for it?), Web 2.0 sites allow, and are even defined by, much more subjective inputs such as personalized messages, photos, and the like. Such things are harder, if not impossible, to quantify.

There are certainly nonhuman technologies involved in Web 2.0—computers, the websites themselves—but human users are, by definition, much more able to manipulate content on Web 2.0 than on Web 1.0. While Web 1.0 sites are centrally conceived, prestructured, and largely immune to manipulation and alteration by users, Web 2.0 sites are based on the whole idea that users can, indeed must, manipulate and alter the sites in innumerable ways. In other words, humans have gained some control over the technologies that on Web 1.0 totally controlled them.

This leads to the issue of the irrationality of rationality and to the conclusion that Web 2.0 serves to reduce or eliminate such irrationalities, especially dehumanization, in comparison to Web 1.0 (to say nothing of irrationalities associated with material realities such as the fast-food restaurant). Web 2.0 is clearly a far more humanized technology than Web 1.0. Indeed, in more fully using the skills and abilities of users, it could be argued that Web 2.0 is a far more "reasonable" system than Web 1.0. User behavior is not highly directed, as is the case with Web 1.0, but is, instead, more creative in nature.

Overall, then, Web 2.0 involves a process of the deMcDonaldization of the Internet, at least in comparison to Web 1.0. In those senses, it contradicts the argument of ever-increasing McDonaldization. Because of user-generated content, Web 2.0 loses something with respect to calculability, efficiency, predictability, and control through nonhuman technologies, but these dimensions, and McDonaldization more generally, have not disappeared completely on Web 2.0. Although content might be personalized and creatively produced on Web 2.0 sites such as eBay and Facebook, McDonaldization continues to exist (as we saw above in the case of eBay) on those sites, especially in their underlying structures. For example, efficiency is manifest in Facebook's largely hidden profit models that are based on the utilization of the creation of value by unpaid prosumers.

Facebook also exerts control, and in fact constitutes an unprecedented intrusion of technology into socializing and selfhood, through the application of nonhuman technologies to these processes. Facebook, for instance, structures social networking by dictating the look of the profiles. Interaction on Facebook follows preset and centrally controlled principles and structures. Examples include writing on someone's "wall" and the constant feed of updates on everything your Facebook "friends" are doing with their profiles. Identity is chosen by selecting from Facebook-determined options and checkboxes, with the result that profile pages look very similar. MySpace, on the other hand, while still very much part of Web 2.0, has lost much market share to Facebook by giving users not expert in Web design the ability to customize and personalize their digital presentations of self to a far greater degree than has Facebook. In other words, the underlying structure of MySpace is not as McDonaldized as that of Facebook. As a result, MySpace profiles are often difficult to navigate. Facebook, however, has much more uniform profiles, where everyone's page has a very similar look. By imposing more structure, Facebook has provided a clean interface that is user-friendly and promotes content development. In structuring the processes of online social networking

and the digital presentation of self, we might argue that, on Facebook, socializing itself has been McDonaldized.

However, this is not to downplay the importance of the customization
allowed by Facebook. While there are ways in which Facebook McDonaldizes
socialization, Facebook is also *de*McDonaldizing the Web experience. Facebook
makes the Web more human through increased social interaction. The experience is highly unpredictable since one is interacting with fellow (unpredictable)
humans. It is difficult, if not impossible, to quantify such interactions; it is their
quality that matters most. Facebook makes keeping in touch with distant
friends efficient. However, it also is highly inefficient since it also facilitates
more socialization through the maintaining of social ties that one might, in the
past, have lost. And it is inefficient because of the huge amount of time and
energy people devote to these sites uploading pictures or socializing online.

One of the things that the preceding discussion makes clear is that while Web
2.0 is not without its McDonaldized elements, it is certainly less McDonaldized
than Web 1.0. Another is that, while Web 2.0 has been discussed as if it is of
one piece, there are important differences among its sites (e.g., Facebook vs.
MySpace).

This analysis of Web 1.0 and Web 2.0 constitutes a kind of test of
McDonaldization as a "grand narrative" and the idea that we are likely to see
an ever-increasing McDonaldization of society. To the degree that Web 2.0 is a
later stage in the development of the Internet than Web 1.0, and of increasing
relevance in society in general, it would be predicted that Web 2.0 would be
more McDonaldized than Web 1.0. That this is not the case, and the fact that
at least on the surface, Web 2.0 is far *less* McDonaldized than Web 1.0, casts
doubt on the McDonaldization thesis.

This leads us to a more nuanced conclusion to this book and to this discussion of McDonaldization and deMcDonaldization. That is, the McDonaldization
process continues apace in the infrastructure of Web 2.0, and it is that which,
paradoxically, has allowed for the deMcDonaldization of the surface-level content to be found there. This is a more complex conclusion than one that simply
depicts a grand narrative of ever escalating and expanding McDonaldization.
This change is not surprising since the McDonaldization thesis was developed
to analyze the material world of fast-food restaurants and the like, but the
digital world—especially the more advanced digital world of Web 2.0—is a
very different place. We should not be shocked that the McDonaldization thesis
requires some emendation in this new context. Indeed, it would have been
shocking had it not required any alteration. In any case, the distinction between
surface-level deMcDonaldization and underlying McDonaldization should
prove useful, especially in analyzing the Internet of the future—Web 3.0, or
whatever that next stage is to be called—as well as the future of society more
generally.

Notes

Chapter 1

1. For a similar but narrower viewpoint to the one expressed here, see Benjamin R. Barber. "Jihad vs. McWorld." *The Atlantic Monthly,* March 1992, pp. 53–63; and also by Barber, *Jihad vs. McWorld.* New York: Times Books, 1995. For a more popular discussion of a similar conflict, see Thomas L. Friedman. *The Lexus and the Olive Tree: Understanding Globalization.* New York: Farrar, Straus, and Giroux, 1999.
2. This stands in contrast to Eric Schlosser's (2001) best-selling *Fast Food Nation* (Boston: Houghton Mifflin), which is about the fast-food industry and devotes much attention to McDonald's.
3. Since the publication of the first edition of this book in 1993, the term *McDonaldization* has, at least to some degree, become part of the academic and public lexicon. For example, among the academic works are Dennis Hayes and Robin Wynyard, eds. *The McDonaldization of Higher Education.* Westport, CT: Bergin and Garvey, 2002; John Drane. *The McDonaldization of the Church.* London: Darton, Longman and Todd, 2001; John Drane. *After McDonaldization: Mission, Ministry, and Christian Discipleship in an Age of Uncertainty.* Grand Rapids, MI: Baker Academic, 2008; Donna Dustin. *The McDonaldization of Social Work.* Farnham, Surrey, UK: Ashgate, 2008; Barry Smart, ed. *Resisting McDonaldization.* London: Sage, 1999; Mark Alfino, John Caputo, and Robin Wynyard, eds. *McDonaldization Revisited.* Westport, CT: Greenwood, 1998; a special issue of the Dutch journal *Sociale Wetenschappen* (vol. 4, 1996) devoted to McDonaldization; the essays in my *McDonaldization: The Reader,* 3rd ed. Thousand Oaks, CA: Sage, 2010; and a special issue (also edited by me) of the *American Behavioral Scientist* titled "McDonaldization: Chicago, America, the World" (October 2003). One also finds many mentions of McDonaldization in the popular media.
4. Alan Bryman has suggested the term *Disneyization,* which he defines in a parallel manner: "the process by which the principles of Disney theme parks are coming to dominate more and more sectors of American society as well as the rest of the world" (p. 26). See Alan Bryman. "The Disneyization of Society." *Sociological Review* 47 (February 1999): 25–47; and Alan Bryman. *The Disneyization of Society.* London: Sage, 2004.
5. See George Ritzer, ed. *McDonaldization: The Reader,* 3rd ed. Thousand Oaks, CA: Sage, 2010.
6. http://aboutmcdonalds.com
7. Ibid.
8. www.weathersealed.com/2009/09/22where-the-buffalo-roamed
9. Martin Plimmer. "This Demi-Paradise: Martin Plimmer Finds Food in the Fast Lane Is Not to His Taste." *Independent* (London), January 3, 1998, p. 46.
10. http://aboutmcdonalds.com
11. McDonald's 2008 Annual Report.

12. www.entrepreneur.com/article/217721
13. International Franchise Association: www.franchise.org
14. In 2008, McDonald's completed the sale of 1,571 company-owned restaurants to a developmental licensee organization. Moreover, in 2008, McDonald's refranchised 675 restaurants, with a goal of refranchising between 1,000 and 1,500 restaurants by 2010. (The rest are either company owned or affiliates; McDonald's 2008 Annual Report.) McDonald's invested in a Denver chain, Chipotle, in 1998 and became its biggest investor in 2001. At the time, Chipotle had 15 stores. By the time McDonald's divested itself of its interest in the company on October 13, 2006, there were more than 500 Chipotle restaurants. In 2008, McDonald's also divested itself from Boston Market, Pret A Manger, and Redbox.
15. Yum! Brands website: www.yum.com/company/ourbrands.asp
16. Yum! Brands 2010 Annual Report: www.yum.com/annualreport/default.asp
17. Subway website: www.world.subway.com
18. Subway press release: "Subway Restaurants Named Number One Franchise." January 2003.
19. Janet Adamy. "For Subway, Anywhere Is Possible Franchise Site." *Wall Street Journal Online*, September 1, 2006.
20. Glenn Collins. "A Big Mac Strategy at Porterhouse Prices." *New York Times*, August 13, 1996, p. D1.
21. Ibid.
22. Ibid.
23. A similarly high-priced chain of steakhouses, Ruth's Chris, claims, perhaps a little too loudly and self-consciously, "Ours is not a McDonald's concept" (Glenn Collins. "A Big Mac Strategy at Porterhouse Prices." *New York Times*, August 13, 1996, p. D1). Even if it is true (and that's doubtful), it makes it clear that all restaurants of this type must attempt to define themselves, either positively or negatively, against the standard set by McDonald's.
24. Morton's 2010 Annual Report: http://investor.mortons.com
25. Timothy Egan. "Big Chains Are Joining Manhattan's Toy Wars." *New York Times*, December 8, 1990, p. 29.
26. Stacey Burling. "Health Club . . . for Kids." *Washington Post*, November 21, 1991, p. D5.
27. Tamar Lewin. "Small Tots, Big Biz." *New York Times Magazine*, January 19, 1989, p. 89.
28. www.curves.com/about-curves/history.php; Lauren L. O'Toole. "McDonald's at the Gym? A Tale of Two Curves." *Qualitative Sociology* 32 (2009): 75–91.
29. Curves Company Facts Sheet 2009.
30. www.curves.com/about-curves/history.php
31. http://aboutmcdonalds.com
32. McDonald's website: www.mcdonalds.com/corp.html
33. www.med-holdings.co.jp/pdf/2009/2009_3rdq_e.pdf
34. www.bloomberg.com/news/2011-01-26/mcdonalds-no-match-for-kfc-in-china-where-colonel-sanders-rules-fast-food.html
35. Mike Comerford. "The Forbidden Kitchen Technology and Testing Help McDonald's Grow Around the Globe." *Chicago Daily Herald-Business*, December 11, 2006; Reuters. "McDonald's Eye 500 Stores in China in 3 Years: Exec."
36. www.bloomberg.com/news/2011–01–26/mcdonalds-no-match-for-kfc-in-china-where-colonel-sanders-rules-fast-food.html
37. http://aboutmcdonalds.com
38. Andrew E. Kramer. "Delivering on Demand: American Fast Food Meets a Warm Reception in Russia." *New York Times*, August 4, 2011, pp. B1, B4.
39. Robin Young. "Britain Is Fast-Food Capital of Europe." *Times* (London), April 25, 1997. However, McDonald's has recently encountered difficulties there.
40. Ilene R. Prusher. "McDonaldized Israel Debates Making Sabbath 'Less Holy.'" *Christian Science Monitor*, January 30, 1998, p. 8; see also, Uri Ram. "Glocommodification: How the Global Consumes the Local McDonald's in Israel." *Current Sociology* 52 (2004): 11–31.

41. http://walmartstores.com/sites/annualreport/2011/financials.aspx; Wal-Mart Corporate Fact Sheet, August 2009.
42. Wal-Mart website: www.walmartstores.com
43. www.timhortons.com/un/en/about/profile.html; Les Whittington. "Tim Hortons: Canada Success Story." *Gazette* (Montreal), October 17, 1997, pp. F4ff.
44. Bloomberg website: www.bloomberg.com/apps/news?pid=10000082&sid=atSds_icml2g&refer=canada, retrieved May 2007.
45. Eric Margolis. "Fast Food: France Fights Back." *Toronto Sun,* January 16, 1997, p. 12.
46. Valerie Reitman. "India Anticipates the Arrival of the Beefless Big Mac." *Wall Street Journal,* October 20, 1993, pp. B1, B3.
47. www.mosburger.com.sg/global_network.php
48. Mos Food Services website: www.mos.co.jp; Mos Burger 2008 Business Report.
49. Alison Leigh Cowan. "Unlikely Spot for Fast Food." *New York Times,* April 29, 1984, sec. 3, p. 5.
50. Peter S. Goodman. "Familiar Logo on Unfamiliar Eateries in Iraq." *Washington Post,* May 26, 2003, pp. A1, A14.
51. www.thebodyshop.com/_en/_ww/services/aboutus_company.aspx;www.thebodyshop-usa.com/beauty/about-us
52. Philip Elmer-Dewitt. "Anita the Agitator." *Time,* January 25, 1993, pp. 52ff; Eben Shapiro. "The Sincerest Form of Rivalry." *New York Times,* October 19, 1991, pp. 35, 46; Bath & Body Works website: www.bathandbodyworks.com
53. Stephanie Clifford. "With an Offbeat Take on Fast-Food Service, Pret A Manger Is Gaining a U.S. Foothold." *New York Times,* 2011, Sunday Business, pp. 1, 5; www.pretamanger.co.uk/philosophy; www.pret.com/about/
54. http://global.campero.com/index.php?cache=1&showPage=38; "Pollo Campero Refreshes Brand Logo Getting Ready for Expansion." *Business Wire,* June 16, 2006.
55. www.jollibee.com.ph/international/usa/store-locator
56. www.pollotropical.com/IntLocations.aspx; Hugh Morley. "A Hunger for the Hispanic: Combining Fast Food, Ethnic Cuisine." *The Record* (Bergen County, NJ), March 22, 2006, p. B01; www.pollotropical.com
57. Lauren Collins. "House Perfect: Is the IKEA Ethos Comfy or Creepy?" *New Yorker,* October 3, 2011, p. 55.
58. Ibid., p. 63.
59. "Stylish, Swedish, 60-ish; Ikea's a Global Phenomenon." *Western Mail,* May 20, 2003, p. 1.
60. "The IKEA Group 2008: Facts and Figures." http://193.108.42.168/repository/documents/1562.pdf
61. Lauren Collins. "House Perfect: Is the IKEA Ethos Comfy or Creepy?" *New Yorker,* October 3, 2011, p. 56.
62. http://about.hm.com
63. H&M website: www.hm.com; H&M 2008 Annual Report.
64. www.inditex.com
65. www.inditex.com, retrieved January 2007; Inditex 2008 Annual Report.
66. Michael Arndt. "McDonald's Goes 24/7," www.msnbc.msn.com/id/ 16828944, retrieved May 2007; McDonald's 2008 Annual Report.
67. Marshall Fishwick, ed. *Ronald Revisited: The World of Ronald McDonald.* Bowling Green, OH: Bowling Green University Press, 1983.
68. John F. Harris. "McMilestone Restaurant Opens Doors in Dale City." *Washington Post,* April 7, 1988, p. D1.
69. E. R. Shipp. "The McBurger Stand That Started It All." *New York Times,* February 27, 1985, sec. 3, p. 3.
70. McDonald's website: www.mcdonalds.com/corp/news/media.html
71. Bill Keller. "Of Famous Arches, Beeg Meks and Rubles." *New York Times,* January 28, 1990, sec. 1, pp. 1, 12.

72. "Wedge of Americana: In Moscow, Pizza Hut Opens 2 Restaurants." *Washington Post,* September 12, 1990, p. B10.
73. Jeb Blount. "Frying Down to Rio." *Washington Post/Business,* May 18, 1994, pp. F1, F5.
74. Thomas L. Friedman. *The Lexus and the Olive Tree: Understanding Globalization.* New York: Farrar, Straus and Giroux, 1999, p. 235.
75. Thomas Friedman. "A Manifesto for the Fast World." *New York Times Magazine,* March 28, 1999, pp. 43–44.
76. *The Economist* website: www.economist.com/blogs/dailychart/2011/07/big-mac-index .html; "Cheesed Off," *The Economist,* February 16, 2009.
77. Reflective of the emergence of a newer global icon, an Australian bank has developed a similarly tongue-in-cheek idea—"the iPod Index" (see http://smh.com.au/news/).
78. "An Alternative Big Mac Index." Economist.com. August 29, 2009; www.economist.com/ daily/chartgallery/displaystory.cfm?story_id=14288808
79. Thomas Friedman. "A Manifesto for the Fast World." *New York Times Magazine,* March 28, 1999, p. 84.
80. Conrad Kottak. "Rituals at McDonald's." In Marshall Fishwick, ed., *Ronald Revisited: The World of Ronald McDonald.* Bowling Green, OH: Bowling Green University Press, 1983, pp. 52–58.
81. Bill Keller. "Of Famous Arches, Beeg Meks and Rubles." *New York Times,* January 28, 1990, sec. 1, pp. 1, 12.
82. William Severini Kowinski. *The Malling of America: An Inside Look at the Great Consumer Paradise.* New York: William Morrow, 1985, p. 218.
83. Stephen M. Fjellman. *Vinyl Leaves: Walt Disney World and America.* Boulder, CO: Westview, 1992.
84. Bob Garfield. "How I Spent (and Spent and Spent) My Disney Vacation." *Washington Post/ Outlook,* July 7, 1991, p. B5. See also Margaret J. King. "Empires of Popular Culture: McDonald's and Disney." In Marshall Fishwick, ed., *Ronald Revisited: The World of Ronald McDonald.* Bowling Green, OH: Bowling Green University Press, 1983, pp. 106–119.
85. Steven Greenhouse. "The Rise and Rise of McDonald's." *New York Times,* June 8, 1986, sec. 3, p. 1.
86. Richard L. Papiernik. "Mac Attack?" *Financial World,* April 12, 1994, p. 30.
87. Laura Shapiro. "Ready for McCatfish?" *Newsweek,* October 15, 1990, pp. 76–77; N. R. Kleinfeld. "Fast Food's Changing Landscape." *New York Times,* April 14, 1985, sec. 3, pp. 1, 6.
88. Gilbert Chan. "Fast Food Chains Pump Profits at Gas Stations." *Fresno Bee,* October 10, 1994, p. F4.
89. Cynthia Rigg. "McDonald's Lean Units Beef Up NY Presence." *Crain's New York Business,* October 31, 1994, p. 1.
90. Anthony Flint. "City Official Balks at Placement of McDonald's at New Courthouse." *Boston Globe,* March 9, 1999, p. B3.
91. Henry Samuel. "McDonald's Restaurants to Open at the Louvre." *Daily Telegraph.* www.telegraph.co.uk/news/worldnews/europe/france/6259044/McDonalds-restaurants-to-open-at-the-Louvre.html
92. Louis Uchitelle. "That's Funny, Those Pickles Don't Look Russian." *New York Times,* February 27, 1992, p. A4.
93. Center for Defense Information website: www.cdi.org/russia
94. Nicholas D. Kristof. "Billions Served (and That Was Without China)." *New York Times,* April 24, 1992, p. A4.
95. Anita Kumar. "A New Food Revolution on Campus." *St. Petersburg Times,* May 11, 2003, p. 1A.
96. Carole Sugarman. "Dining Out on Campus." *Washington Post/Health,* February 14, 1995, p. 20.
97. Edwin McDowell. "Fast Food Fills Menu for Many Hotel Chains." *New York Times,* January 9, 1992, pp. D1, D6.

98. Dan Freedman. "Low Fat? The Kids Aren't Buying; Districts Struggle to Balance Mandates for Good Nutrition With Reality in the Cafeteria." *The Times Union,* September 22, 2002, p. A1.

99. "Back to School: School Lunches." *Consumer Reports,* September 1998, p. 49.

100. Mike Berry. "Redoing School Cafeterias to Favor Fast-Food Eateries." *Orlando Sentinel,* January 12, 1995, p. 11.

101. "Pediatric Obesity: Fast-Food Restaurants Cluster Around Schools." *Obesity, Fitness and Wellness Week,* September 24, 2005, p. 1517.

102. "Grade 'A' Burgers." *New York Times,* April 13, 1986, pp. 12, 15.

103. Jennifer Curtis. "McDonald's Attacked for Toys That Push Its Fatty Fast Food." *The West Australian* (Perth), January 16, 2007, p. 13.

104. Lindsey Tanner. "Pediatric Hospitals That Serve Fast Food Raise More Alarm." *Houston Chronicle,* December 28, 2006, p. B2.

105. Gloria Pitzer. *Secret Fast Food Recipes: The Fast Food Cookbook.* Marysville, MI: Author.

106. This discussion is derived from George Ritzer. "Revolutionizing the World of Consumption." *Journal of Consumer Culture* 2 (2002): 103–118.

107. George Anders. "McDonald's Methods Come to Medicine as Chains Acquire Physicians' Practices." *Wall Street Journal,* August 24, 1993, pp. B1, B6.

108. Peter Prichard. *The Making of McPaper: The Inside Story of* USA TODAY. Kansas City, MO: Andrews, McMeel and Parker, 1987.

109. I would like to thank Lee Martin for bringing this case (and menu) to my attention.

110. Peter Prichard. *The Making of McPaper: The Inside Story of* USA TODAY. Kansas City, MO: Andrews, McMeel and Parker, 1987, pp. 232–233.

111. Howard Kurtz. "Slicing, Dicing News to Attract the Young." *Washington Post,* January 6, 1991, p. Al.

112. Kathryn Hausbeck and Barbara G. Brents. "McDonaldization of the Sex Industries? The Business of Sex." In George Ritzer, ed. *McDonaldization: The Reader,* 3rd ed. Thousand Oaks, CA: Sage, 2010, pp. 102–117.

113. Martin Gottlieb. "Pornography's Plight Hits Times Square." *New York Times,* October 5, 1986, sec. 3, p. 6.

114. http://sociologycompass.wordpress.com/2009/11/02/augmented-reality-going-the-way-of-the-dildo/

115. Jean Sonmor. "Can We Talk Sex: Phone Sex Is Hot-Wiring Metro's Lonely Hearts." *Toronto Sun,* January 29, 1995, pp. M11ff.

116. Ibid.

117. For a selection of this work, see George Ritzer, ed. *McDonaldization: The Reader,* 3rd ed. Thousand Oaks, CA: Sage, 2010.

118. Ian Heywood, "Urgent Dreams: Climbing, Rationalization, and Ambivalence." In George Ritzer, ed. *McDonaldization: The Reader,* 3rd ed. Thousand Oaks, CA: Sage, 2010, pp. 65–69.

119. Matthew B. Robinson, "McDonaldization of America's Police, Courts, and Corrections." In George Ritzer, ed. *McDonaldization: The Reader,* 3rd ed. Thousand Oaks, CA: Sage, 2010, pp. 85–100.

120. Sara Raley. "McDonaldization and the Family." In George Ritzer, ed. *McDonaldization: The Reader,* 3rd ed. Thousand Oaks, CA: Sage, 2010, pp. 138–148.

121. Gary Wilkinson. "McSchools for McWorld: Mediating Global Pressures With a McDonaldizing Education Policy Response." In George Ritzer, ed. *McDonaldization: The Reader,* 3rd ed. Thousand Oaks, CA: Sage, 2010, pp. 150–157.

122. Lee F. Monaghan. "McDonaldizing Men's Bodies? Slimming, Associated (Ir)Rationalities and Resistances." In George Ritzer, ed. *McDonaldization: The Reader,* 3rd ed. Thousand Oaks, CA: Sage, 2010, pp. 119–136.

123. Nathan Jurgenson. "The De-McDonaldization of the Internet." In George Ritzer, ed. *McDonaldization: The Reader,* 3rd ed. Thousand Oaks, CA: Sage, 2010, pp. 159–170.

124. Andrew J. Knight. "Supersizing Farms: The McDonaldization of Agriculture." In George Ritzer, ed. *McDonaldization: The Reader,* 3rd ed. Thousand Oaks, CA: Sage, 2010, pp. 192–205.

125. John Drane. "From Creeds to Burgers: Religious Control, Spiritual Search, and the Future of the World." In George Ritzer, ed. *McDonaldization: The Reader*, 3rd ed. Thousand Oaks, CA: Sage, 2010, pp. 222–227.

126. Jos Gamble. "Multinational Retailers in China: Proliferating 'McJobs' or Developing Skills?" In George Ritzer, ed. *McDonaldization: The Reader*, 3rd ed. Thousand Oaks, CA: Sage, 2010, pp. 172–190.

127. Bryan Turner. "McCitizens: Risk, Coolness and Irony in Contemporary Policy." In George Ritzer, ed. *McDonaldization: The Reader*, 3rd ed. Thousand Oaks, CA: Sage, 2010, pp. 229–232.

128. Arthur Asa Berger. *Signs in Contemporary Culture: An Introduction to Semiotics*, 2nd ed. Salem, WI: Sheffield, 1999.

129. Max Weber. *Economy and Society*. Totowa, NJ: Bedminster, 1921/1968; Stephen Kalberg. "Max Weber's Types of Rationality: Cornerstones for the Analysis of Rationalization Processes in History." *American Journal of Sociology* 85 (1980): 1145–1179.

130. Ian Mitroff and Warren Bennis. *The Unreality Industry: The Deliberate Manufacturing of Falsehood and What It Is Doing to Our Lives*. New York: Birch Lane, 1989, p. 142.

131. Melanie Warner. "McDonald's Revival Has Hidden Health Costs." *International Herald Tribune*, April 20, 2006, p. 17.

132. Melanie Warner. "U.S. Restaurant Chains Find There Is No Too Much." *New York Times*, July 28, 2006, p. C5.

133. Martin Plimmer. "This Demi-Paradise: Martin Plimmer Finds Food in the Fast Lane Is Not to His Taste." *Independent* (London), January 3, 1998, p. 46.

134. As we will see in Chapter 4, this increased control often comes from the substitution of nonhuman for human technology.

135. It should be pointed out that the words *rational, rationality,* and *rationalization* are being used differently here and throughout the book than they are ordinarily employed. For one thing, people usually think of these terms as being largely positive; something that is rational is usually considered good. However, they are used here in a generally negative way. The positive term in this analysis is genuinely human "reason" (for example, the ability to act and work creatively), which is seen as being denied by inhuman, rational systems such as the fast-food restaurant. For another, the term *rationalization* is usually associated with Freudian theory as a way of explaining away some behavior, but here it describes the increasing pervasiveness of rationality throughout society. Thus, in reading this book, you must be careful to interpret the terms in these ways rather than in the ways they are conventionally employed.

136. Alan Riding. "Only the French Elite Scorn Mickey's Debut." *New York Times*, April 13, 1992, p. A13.

137. George Stauth and Bryan S. Turner. "Nostalgia, Postmodernism and the Critique of Mass Culture." *Theory, Culture and Society* 5 (1988): 509–526; Bryan S. Turner. "A Note on Nostalgia." *Theory, Culture and Society* 4 (1987): 147–156.

138. Lee Hockstader. "No Service, No Smile, Little Sauce." *Washington Post*, August 5, 1991, p. A12.

139. Douglas Farah. "Cuban Fast Food Joints Are Quick Way for Government to Rally Economy." *Washington Post*, January 24, 1995, p. A14.

140. In this sense, this resembles Marx's critique of capitalism. Marx was not animated by a romanticization of precapitalist society but, rather, by the desire to produce a truly human (communist) society on the base provided by capitalism. Despite this specific affinity to Marxist theory, this book is, as you will see, premised far more on the theories of Max Weber.

141. These concepts are associated with the work of the social theorist, Anthony Giddens. See, for example, *The Constitution of Society*. Berkeley: University of California Press, 1984.

142. Jon Ortiz. "Customers Drawn to Ikea Experience." *Sacramento Bee*, February 26, 2006, p. D1ff.

143. Tod Hartman. "On the Ikeaization of France." *Public Culture* 19 (2007): 483–498.

144. Stephanie Irwin. "It's a Destination . . . It's a Lifestyle . . . It's a Furniture Store." *Dayton Daily News*, August 2, 2006, p. D8.

145. Lauren Collins. "House Perfect: Is the IKEA Ethos Comfy or Creepy?" *New Yorker*, October 3, 2011, pp. 54–65.

146. Stephanie Irwin. "It's a Destination . . . It's a Lifestyle . . . It's a Furniture Store." *Dayton Daily News,* August 2, 2006, p. D8.

147. Jon Ortiz. "Customers Drawn to Ikea Experience." *Sacramento Bee,* February 26, 2006, p. D1ff.

148. Ibid.

149. Lauren Collins. "House Perfect: Is the IKEA Ethos Comfy or Creepy?" *New Yorker,* October 3, 2011, p. 57.

150. Stephanie Irwin. "It's a Destination . . . It's a Lifestyle . . . It's a Furniture Store." *Dayton Daily News,* August 2, 2006, p. D8.

151. Robert J. Samuelson. "In Praise of McDonald's." *Washington Post,* November 1, 1989, p. A25.

152. Edwin M. Reingold. "America's Hamburger Helper." *Time,* June 29, 1992, pp. 66–67.

153. I would like to thank my colleague, Stan Presser, for suggesting that I enumerate the kinds of advantages listed on these pages.

154. George Ritzer and Seth Ovadia. "The Process of McDonaldization Is Not Uniform, nor Are Its Settings, Consumers, or the Consumption of Its Goods and Services." In Mark Gottdiener, ed. *New Forms of Consumption: Consumers, Cultures and Commodification.* Lanham, MD: Rowman and Littlefield, 2000, pp. 33–49.

155. Meredith Hoffman. "Battling for Street Fare Honors." *New York Times,* September 20, 2011.

156. Stacy Perman. *In-N-Out Burger.* New York: Collins Business, 2009, p. 26.

157. Ibid., p. 42.

158. Ibid., p. 39.

159. Ibid., p. 96.

160. Ibid., p. 89.

161. Ibid., p. 127.

162. Ibid., p. 129.

163. Ibid., p. 139.

164. Ibid., p. 140.

165. Ibid., p. 242; Tom McNichol. "The Secret Behind Burger Cult." *New York Times,* August 14, 2002.

166. Stacy Perman. *In-N-Out Burger.* New York: Collins Business, 2009, p. 168.

167. Stephanie Clifford. "Would You Like a Smile With That?" *New York Times,* August 6, 2011.

168. www.pret.com/us/about_our_company/about.htm

169. Stephanie Clifford. "Would You Like a Smile With That?" *New York Times,* August 6, 2011.

170. Ibid.

171. Ibid.

172. Ibid.

173. Suzanne Kapner. "Business; From a British Chain, Lunch in a New York Minute." *New York Times,* July 29, 2001.

174. Stephanie Clifford. "Would You Like a Smile With That?" *New York Times,* August 6, 2011.

175. www.dailymail.co.uk/femail/article-1375525/Pret-A-Manger-The-alarming-truth-fresh-healthy-lunch.html#ixzz1Yt2SeKX3

Chapter 2

1. Although the precursors discussed in this chapter do not exhaust the rationalized institutions that predate McDonald's, they are the most important for understanding McDonald's and McDonaldization.

2. This discussion of Weber's ideas is based on Max Weber. *Economy and Society.* Totowa, NJ: Bedminster, 1921/1968.

3. The fast-food restaurant can also be seen as part of a bureaucratic system; in fact, huge conglomerates (for example, Yum! Brands, Inc.) now own many of the fast-food chains.

4. Weber called the latter substantive rationality, to distinguish it from formal rationality.

5. Ronald Takaki. *Iron Cages: Race and Culture in 19th-Century America.* New York: Oxford University Press, 1990, p. ix.

6. Harvey Greisman. "Disenchantment of the World." *British Journal of Sociology* 27 (1976): 497–506.

7. For a discussion of cruise ships as "cathedrals of consumption," see George Ritzer. *Enchanting a Disenchanted World: Revolutionizing the Means of Consumption,* 3rd ed. Thousand Oaks, CA: Sage, 2010.

8. Zygmunt Bauman. *Modernity and the Holocaust.* Ithaca, NY: Cornell University Press, 1989, p. 149.

9. Ibid., p. 8.

10. However, in contemporary Rwanda, an estimated 800,000 people were killed in 100 days (three times the rate of Jewish dead during the Holocaust) in warfare between the Hutus and Tutsis. The methods employed—largely machete—were decidedly not rationalized. See Philip Gourevitch. *We Wish to Inform You That Tomorrow We Will Be Killed With Our Families: Stories From Rwanda.* New York: Farrar, Straus and Giroux, 1998.

11. As you will see in Chapter 3, the fast-food restaurants enhance their efficiency by getting customers to perform (without pay) a variety of their tasks.

12. Zygmunt Bauman. *Modernity and the Holocaust.* Ithaca, NY: Cornell University Press, 1989, p. 103.

13. Ibid., p. 89.

14. Ibid., p. 8.

15. Ibid., p. 102.

16. Feingold, cited in Zygmunt Bauman. *Modernity and the Holocaust.* Ithaca, NY: Cornell University Press, 1989, p. 136.

17. Frederick W. Taylor. *The Principles of Scientific Management.* New York: Harper & Row, 1947; Robert Kanigel. *One Best Way: Frederick Winslow Taylor and the Enigma of Efficiency.* New York: Viking, 1997.

18. Frederick W. Taylor. *The Principles of Scientific Management.* New York: Harper & Row, 1947, pp. 6–7.

19. Ibid., p. 11.

20. Henry Ford. *My Life and Work.* Garden City, NY: Doubleday, 1922; James T. Flink. *The Automobile Age.* Cambridge: MIT Press, 1988.

21. Henry Ford. *My Life and Work.* Garden City, NY: Doubleday, 1922, p. 80.

22. Jerry Newman. *My Secret Life on the McJob.* New York: McGraw-Hill, 2007, pp. 168–169.

23. Bruce A. Lohof. "Hamburger Stand Industrialization and the Fast-Food Phenomenon." In Marshall Fishwick, ed., *Ronald Revisited: The World of Ronald McDonald.* Bowling Green, OH: Bowling Green University Press, 1983, p. 30; see also Ester Reiter. *Making Fast Food.* Montreal: McGill-Queen's University Press, 1991, p. 75.

24. Marshall Fishwick. "Cloning Clowns: Some Final Thoughts." In Marshall Fishwick, ed., *Ronald Revisited: The World of Ronald McDonald.* Bowling Green, OH: Bowling Green University Press, 1983, pp. 148–151. For more on the relationship described in the same paragraph between the automobile and the growth of the tourist industry, see James T. Flink. *The Automobile Age.* Cambridge: MIT Press, 1988.

25. General Motors, especially Alfred Sloan, further rationalized the automobile industry's bureaucratic structure. Sloan is famous for GM's multidivisional system, in which the central office handled long-range decisions while the divisions made the day-to-day decisions. This innovation proved so successful in its day that the other automobile companies as well as many other corporations adopted it. See James T. Flink. *The Automobile Age.* Cambridge: MIT Press, 1988; Alfred P. Sloan Jr. *My Years at General Motors.* Garden City, NY: Doubleday, 1964.

26. "Levitt's Progress." *Fortune,* October 1952, pp. 155ff.

27. Richard Perez-Pena. "William Levitt, 86, Suburb Maker, Dies." *New York Times,* January 29, 1994, p. 26.

28. "The Most House for the Money." *Fortune,* October 1952, p. 152.

29. Ibid., p. 153.
30. Herbert Gans. *The Levittowners: Ways of Life and Politics in a New Suburban Community.* New York: Pantheon, 1967, p. 13.
31. Patricia Dane Rogers. "Building . . ." *Washington Post/Home,* February 2, 1995, pp. 12, 15; Rebecca Lowell. "Modular Homes Move Up." *Wall Street Journal,* October 23, 1998, p. W10.
32. Richard E. Gordon, Katherine K. Gordon, and Max Gunther. *The Split-Level Trap.* New York: Gilbert Geis Associates, 1960.
33. Georgia Dullea. "The Tract House as Landmark." *New York Times,* October 17, 1991, pp. C1, C8.
34. Herbert Gans. *The Levittowners: Ways of Life and Politics in a New Suburban Community.* New York: Pantheon, 1967, p. 432.
35. William Severini Kowinski. *The Malling of America: An Inside Look at the Great Consumer Paradise.* New York: William Morrow, 1985.
36. www.forbes.com/business/2007/01/09/malls-world-largest-biz-cx_tvr_0109malls.ht
37. http://theseoultimes.com/ST/db/read.php?idx=1962
38. www.easternct.edu/depts/amerst/MallsWorld.htm
39. Janice L. Kaplan. "The Mall Outlet for Cabin Fever." *Washington Post/Weekend,* February 10, 1995, p. 53.
40. William Severini Kowinski. *The Malling of America: An Inside Look at the Great Consumer Paradise.* New York: William Morrow, 1985, p. 25. For a discussion of the significance of the mall in the history of consumption, see Lizabeth Cohen. *Consumer's Republic: The Politics of Mass Consumption in Postwar America.* New York: Knopf, 2003, especially Chapter 6.
41. http://cannmoney.printthis.clickability.com/pt/cpt?action=cpt&title=dead+ stores+and+mall
42. David Segal. "Our Love Affair With Malls Is on the Rocks." *New York Times,* January 31, 2008, pp. 1 Businessff.
43. http://cnnmoney.printthis.clickability.com/pt/cpt?action=cpt&title=Credit+ crisis%2C+spec
44. Ray Kroc. *Grinding It Out.* New York: Berkeley Medallion Books, 1977; Stan Luxenberg. *Roadside Empires: How the Chains Franchised America.* New York: Viking, 1985; and John F. Love. *McDonald's: Behind the Arches.* Toronto: Bantam, 1986.
45. John F. Love. *McDonald's: Behind the Arches.* Toronto: Bantam, 1986, p. 18.
46. Ibid., p. 20.
47. Thomas S. Dicke. *Franchising in America: The Development of a Business Method, 1840–1980.* Chapel Hill: University of North Carolina Press, 1992, pp. 2–3.
48. Taco Bell website: www.tacobell.com
49. John Vidal. *McLibel: Burger Culture on Trial.* New York: New Press, 1997, p. 34.
50. Wayne Huizenga played a similar role in the video business by taking over a chain developed by a Dallas entrepreneur and turning it into the Blockbuster empire. See David Altaner. "Blockbuster Video: 10 Years Running Family-Oriented Concept Has Changed Little Since 1985, When Chain Was Founded by a Dallas Businessman." *Sun-Sentinel* (Fort Lauderdale, FL), October 16, 1995, pp. 16ff.
51. John F. Love. *McDonald's: Behind the Arches.* Toronto: Bantam, 1986, pp. 68–69.
52. McDonald's website: www.mcdonalds.com/corp.html
53. Like McDonald's Hamburger University, Burger King set up its own Burger King University in 1978; see Ester Reiter. *Making Fast Food.* Montreal: McGill-Queen's University Press, 1991, p. 68.
54. John F. Love. *McDonald's: Behind the Arches.* Toronto: Bantam, 1986, pp. 141–142.
55. Mark Bittman. "Is Junk Food Really Cheaper?" *New York Times–Sunday Review,* September 21, 2011, pp. 1, 7.
56. Joe Kincheloe. "The Complex Politics of McDonald's and the New Childhood: Colonizing Kidworld." In Gaile S. Cannella and Joe L. Kincheloe, eds., *Kidworld: Childhood Studies, Global Perspectives, and Education.* New York: Peter Lang, 2002, pp. 75–122.
57. Ironically and paradoxically, some aspects of the process of McDonaldization (for example, the Internet and cybershops) are allowing many people to do more things at home. This poses something of a threat to other rationalized aspects of society (for example, shopping malls).

58. Ester Reiter. *Making Fast Food*. Montreal: McGill-Queen's University Press, 1991, p. 165.

59. Don Slater. "'You Press the Button, We Do the Rest': Some Thoughts on the McDonaldization of the Internet." Paper presented at the meetings of the Eastern Sociological Society, Boston, March 1999.

60. Daniel Bell. *The Coming of Post-Industrial Society: A Venture in Social Forecasting*. New York: Basic Books, 1973.

61. Jerald Hage and Charles H. Powers. *Post-Industrial Lives: Roles and Relationships in the 21st Century*. Newbury Park, CA: Sage, 1992.

62. Ibid., p. 10.

63. Although there are, as we have seen, efforts to automate them as well.

64. Jerald Hage and Charles H. Powers. *Post-Industrial Lives: Roles and Relationships in the 21st Century*. Newbury Park, CA: Sage, 1992, p. 50.

65. Pierre Bourdieu. *Distinction: A Social Critique of the Judgment of Taste*. Cambridge, MA: Harvard University Press, 1984.

66. For more on postmodernism, see George Ritzer. *Postmodern Social Theory*. New York: McGraw-Hill, 1997; Jean Baudrillard. *Symbolic Exchange and Death*. London: Sage, 1976/1993; Fredric Jameson. "Postmodernism, or the Cultural Logic of Late Capitalism." *New Left Review* 146 (1984): 53–92; Fredric Jameson. *Postmodernism, or The Cultural Logic of Late Capitalism*. Durham, NC: Duke University Press, 1991; Jean-François Lyotard. *The Postmodern Condition: A Report on Knowledge*. Minneapolis: University of Minnesota Press, 1984; Steven Best and Douglas Kellner. *Postmodern Theory: Critical Interrogations*. New York: Guilford, 1991.

67. Smart argues that, rather than viewing modernism and postmodernism as epochs, people can see them as engaged in a long-running and ongoing set of relationships, with postmodernity continually pointing out the limitations of modernity. See Barry Smart. *Postmodernity*. London: Routledge, 1993.

68. David Harvey. *The Condition of Postmodernity: An Enquiry Into the Origins of Cultural Change*. Oxford, UK: Basil Blackwell, 1989, p. 189.

69. Fredric Jameson. "Postmodernism, or the Cultural Logic of Late Capitalism." *New Left Review* 146 (1984): 53–92; Fredric Jameson. *Postmodernism, or the Cultural Logic of Late Capitalism*. Durham, NC: Duke University Press, 1991.

70. Fredric Jameson. "Postmodernism, or the Cultural Logic of Late Capitalism." *New Left Review* 146 (1984): 64.

71. Ian Heywood. "Urgent Dreams: Climbing, Rationalization and Ambivalence." *Leisure Studies* 13 (1994): 179–194.

72. Jon Krakauer. *Into Thin Air*. New York: Anchor, 1997, p. xvii.

73. Ibid., pp. 39, 353.

74. Ibid., p. 320.

75. Ibid.

76. Ibid., p. 100.

77. Ibid., p. 86.

78. Ibid.

79. Sean Patrick Farrell. "New Toll for Mountain Climbers: A Stopwatch." *New York Times*, October 16, 2011, p. 18.

80. Ibid.

81. Ibid.

Chapter 3

1. Herbert Simon. *Administrative Behavior*, 2nd ed. New York: Free Press, 1957.

2. Ray Kroc. *Grinding It Out*. New York: Berkeley Medallion Books, 1977, p. 8.

3. Max Boas and Steve Chain. *Big Mac: The Unauthorized Story of McDonald's*. New York: E. P. Dutton, 1976, pp. 9–10.

4. Ibid.

5. Ray Kroc. *Grinding It Out.* New York: Berkeley Medallion Books, 1977, pp. 96–97.

6. www.washingtonpost.com/lifestyle/magazine/whos-lovin-it/20

7. Jerry Newman. *My Secret Life on the McJob: Lessons From Behind the Counter Guaranteed to Supersize Any Management Style.* New York: McGraw-Hill, 2006, p. 53.

8. Jill Lawrence. "80 Pizzas per Hour." *Washington Post,* June 9, 1996, pp. W07ff.

9. Arthur Kroker, Marilouise Kroker, and David Cook. *Panic Encyclopedia: The Definitive Guide to the Postmodern Scene.* New York: St. Martin's, 1989, p. 119.

10. Michael Lev. "Raising Fast Food's Speed Limit." *Washington Post,* August 7, 1991, p. D1.

11. www.investorworkplace.com/2011/05/mcdonalds/nyse-mcd-touch-screen-menu-ordering/

12. Jim Kershner. "Trays of Our Lives: Fifty Years After Swanson Unveiled the First TV Dinner, Meals-in-a-Box Have Never Been Bigger." *Spokesman Review,* March 19, 2003, p. D1.

13. "The Microwave Cooks Up a New Way of Life." *Wall Street Journal,* September 19, 1989, p. B1; "Microwavable Foods: Industry's Response to Consumer Demands for Convenience." *Food Technology* 41 (1987): 52–63.

14. "Microwavable Foods: Industry's Response to Consumer Demands for Convenience." *Food Technology* 41 (1987): 54.

15. Eben Shapiro. "A Page From Fast Food's Menu." *New York Times,* October 14, 1991, pp. D1, D3.

16. I would like to thank Dora Giemza for the insights into NutriSystem. See also "Big People, Big Business: The Overweight Numbers Rising, Try NutriSystem." *Washington Post/Health,* October 10, 1989, p. 8.

17. Lisa Schnirring. "What's Behind the Women-Only Fitness Center Boom?" *Physician and Sports Medicine* 30 (November 2002): 15.

18. William Severini Kowinski. *The Malling of America: An Inside Look at the Great Consumer Paradise.* New York: William Morrow, 1985, p. 61.

19. www.7-eleven.com/AboutUs/InternationalLicensing/tabid/115/Default.aspx

20. BrewThru website: www.brewthru.com

21. Wendy Tanaka. "Catalogs Deck Halls to Tune of Billions: Mail Order Called 'Necessity' for Consumers." *Arizona Republic,* December 9, 1997, p. A3.

22. www.amazon.com/gp/help/customer/display.html/ref=help_search_1-5?ie=UTF8&nodeId= 14101911&qid=1258841029&sr=1-5

23. http://phx.corporate-ir.net/phoenix.zhtml?c=176060&p=irol-newsArticles&ID=1565581& highlight=

24. Robin Herman. "Drugstore on the Net." *Washington Post/Health,* May 4, 1999, pp. 15ff.

25. Doris Hajewski. "Employees Save Time by Shopping Online at Work." *Milwaukee Journal Sentinel,* December 16, 1998, pp. B1ff.

26. Bruno Giussani. "This Development Is One for the Books." *Chicago Tribune,* September 22, 1998, pp. C3ff.

27. L. Walker. "Google Turns Its Gaze on Online Shopping." *Washington Post,* December, 15, 2002, p. H7.

28. Dennis Hayes and Robin Wynyard, eds. *The McDonaldization of Higher Education.* Westport, CT: Bergin & Garvey, 2002; Martin Parker and David Jary. "The McUniversity: Organization, Management and Academic Subjectivity." *Organization* 2 (1995): 1–19.

29. Linda Perlstein. "Software's Essay Test: Should It Be Grading?" *Washington Post,* October 13, 1998, pp. A1ff.

30. See www.wisetermpapers.com or www.12000papers.com

31. See www.turnitin.com

32. George Ritzer and David Walczak. "The Changing Nature of American Medicine." *Journal of American Culture* 9 (1987): 43–51.

33. Julia Wallace. "Dr. Denton Cooley: Star of 'The Heart Surgery Factory.'" *Washington Post,* July 19, 1980, p. A6.

34. www.eng.cvz.ru

35. "Moving Right Along." *Time,* July 1, 1985, p. 44.

36. www.teleroboticsurgeons.com/davinci.htm; retrieved March 31, 2007.

37. www.drwalkin.com/home

38. http://retailtrafficmag/retailing/operations/drugstore_chains_walk_in_clinics/;   http://minuteclinic.com/en/USA/

39. Mark Potts. "Blockbuster Struggles With Merger Script." *Washington Post/Washington Business,* December 9, 1991, p. 24; Eben Shapiro. "Market Place: A Mixed Outlook for Blockbuster." *New York Times,* February 21, 1992, p. D6.

40. www.hulu.com/about

41. Stephen Fjellman. *Vinyl Leaves: Walt Disney World and America.* Boulder, CO: Westview, 1992.

42. Michael Harrington. "To the Disney Station." *Harper's,* January 1979, pp. 35–39.

43. Lynn Darling. "On the Inside at Parks à la Disney." *Washington Post,* August 28, 1978, p. A10.

44. www.eHarmony.com; www.match.com

45. www.adultfriendfinder.com; www.newyork.craigslist.org/cas/

46. I would like to thank Steve Lankenau for suggesting to me some of the points about McDonaldization and health clubs made here.

47. On another dimension of McDonaldization, exercise machines also offer a high degree of calculability, with many of them registering miles run, level of difficulty, and calories burned.

48. Jeffrey Hadden and Charles E. Swann. *Primetime Preachers: The Rising Power of Televangelism.* Reading, MA: Addison Wesley, 1981.

49. John Drane. *The McDonaldization of the Church.* London: Darton, Longman, and Todd, 2001, p. 36.

50. Don Slater. "'You Press the Button, We Do the Rest': Some Thoughts on the McDonaldization of the Internet." Paper presented at the meetings of the Eastern Sociological Society, Boston, March 6, 1999.

51. JoAnna Daemmrich. "Candidates Increasingly Turn to Internet." *Baltimore Sun,* October, 21, 1998, pp. 1Bff.

52. Glenn Kessler and James Rowell. "Virtual Medical Symposia: Communicating Globally, Quickly, and Economically; Use Internet." *Medical Marketing and Media,* September 1998, pp. 60ff.

53. Russell Blinch. "Instant Message Programs Keep Millions Ecstatic." *Denver Rocky Mountain News,* May 11, 1998, p. 6B.

54. Jennifer Lenhart. "'Happy Holidays,' High-Tech Style." *Washington Post,* December 20, 1998, pp. B1ff.

55. They have already McDonaldized the process of breeding, raising, and slaughtering chickens (see Chapter 4).

56. Janet Adamy. "For McDonald's, It's a Wrap." *Wall Street Journal,* January 30, 2007, pp. B1, B2.

57. Henry Ford. *My Life and Work.* Garden City, NY: Doubleday, 1922, p. 72.

58. www.hrblock.com/presscenter/about/history.jsp

59. George Ritzer and Nathan Jurgenson. "Production, Consumption, Prosumption: The Nature of Capitalism in the Age of the Digital 'Prosumer.'" *Journal of Consumer Culture* 10 (2010): 13–36; George Ritzer, Paul Dean, and Nathan Jurgenson. "The Coming of Age of the Prosumer" (Special issue). *American Behavioral Scientist,* forthcoming, 2012; Daniel Bell. *The Coming of Post-Industrial Society: A Venture in Social Forecasting.* New York: Basic Books, 1973.

60. Thomas R. Ide and Arthur J. Cordell. "Automating Work." *Society* 31 (1994): 68.

61. www.shakensteakfranchise.com/national-expansion-steak-franchise-development.php

62. Steak 'n Shake just announced a merger with the Western Sizzlin Corporation: www.steaknshake.com/investing%5Cnews%5C10222009.pdf

63. www.sweettomatoes.com/ourcompany/

64. www.souplantation.com/pressroom/companyfacts.asp

65. www.fudddruckers.com/franchising/

66. This once-thriving chain of more than 600 restaurants has been reduced to a small number of independent operations in the Northeast; see Sandra Evans. "Roy Rogers Owners Hope for Happy Trails." *Washington Post,* August 4, 1997, pp. F05ff; see www.royrogersrestaurants.com
67. See www.supermarketguru.com/page.cfm/25603
68. www.nextepsystems.com/Home/tabid/36/Default.aspx
69. Eric Palmer. "Scan-do Attitude: Self-Service Technology Speeds Up Grocery Shopping." *Kansas City Star,* April 8, 1998, pp. B1ff.
70. Retail Banking Research news release regarding the "Global EPOS and Self-Checkout 2009."
71. Eben Shapiro. "Ready, Set, Scan That Melon." *New York Times,* June 14, 1990, pp. D1, D8.
72. Ibid.
73. Chris Woodyard. "Grocery Shoppers Can Be Own Cashiers." *USA TODAY,* March 9, 1998, p. 6B.
74. Robert Kisabeth, Anne C. Pontius, Bernard E. Statland, and Charlotte Galper. "Promises and Pitfalls of Home Test Devices." *Patient Care* 31(October 15, 1997): 125ff.
75. Barry Meier. "Need a Teller? Chicago Bank Plans a Fee." *Washington Post,* April 27, 1995, pp. D1, D23.
76. James Barron. "Please Press 2 for Service; Press ? for an Actual Human." *New York Times,* February 17, 1989, pp. A1, B2.
77. Michael Schrage. "Calling the Technology of Voice Mail Into Question." *Washington Post,* October 19, 1990, p. F3.
78. www.2010census.gov/2010census/take10map/
79. Personal communication between Mike Ryan (my assistant) with Rose Cowan at the Bureau of the Census.
80. Just as quality is equated with quantity, quality is also equated with other aspects of McDonaldization, such as "standardization and predictability." See Ester Reiter. *Making Fast Food.* Montreal: McGill-Queen's University Press, 1991, p. 107.
81. Shoshana Zuboff. *In the Age of the Smart Machine: The Future of Work and Power.* New York: Basic Books, 1988.
82. Bruce Horovitz. "Fast-Food Chains Bank on Bigger-Is-Better Mentality." *USA TODAY,* September 12, 1997, p. 1b.
83. In addition, as you will see in Chapter 6, protests against these garish signs helped lead to their virtual disappearance.
84. "Taco Bell Delivers Even Greater Value to Its Customers by Introducing Big Fill Menu." *Business Wire,* November 2, 1994.
85. Melanie Warner. "U.S. Restaurant Chains Find There Is No Too Much." *New York Times,* July 28, 2006, p. C5.
86. Ibid.
87. Philip Elmer-DeWitt. "Fat Times." *Time,* January 16, 1995, pp. 60–65.
88. Jane Wells. "Supersizing It: McDonald's Tests Bigger Burger." www.msnbc.msn.com/id/17757931/fromET, retrieved May 2007.
89. Barbara W. Tuchman. "The Decline of Quality." *New York Times Magazine,* November 2, 1980, p. 38. For example, United Airlines does not tell people anything about the quality of their numerous flights, such as the likelihood that their planes will be on time.
90. Marion Clark. "Arches of Triumph." *Washington Post/Book World,* June 5, 1977, p. G6.
91. A. A. Berger. "Berger vs. Burger: A Personal Encounter." In Marshall Fishwick, ed., *Ronald Revisited: The World of Ronald McDonald.* Bowling Green, OH: Bowling Green University Press, 1983, p. 126.
92. Max Boas and Steven Chain. *Big Mac: The Unauthorized Story of McDonald's.* New York: Dutton, 1976, p. 121.
93. Ibid., p. 117.
94. A. C. Stevens. "Family Meals: Olive Garden Defines Mediocrity." *Boston Herald,* March 2, 1997, p. 055.
95. "The Cheesecake Factory Restaurants Celebrate 25th Anniversary." *Business Wire,* February 25, 2003.

96. "Weekly Prompts From a Mentor." *New York Times,* August 21, 2011, p. A18.
97. "A Way to Speed the Pace." *New York Times,* August 21, 2011, p. A18.
98. Susan Gervasi. "The Credentials Epidemic." *Washington Post,* August 30, 1990, p. D5.
99. Iver Peterson. "Let That Be a Lesson: Rutgers Bumps a Well-Liked but Little-Published Professor." *New York Times,* May 9, 1995, p. B1.
100. www.sciplore.org/publications/2009-Google_Scholar%27s_Ranking_ Algorithm_—_An_ Introductory_Overview_—_preprint.pdf
101. http://code.google.com/p/citations-gadget/
102. Ibid.
103. Kenneth Cooper. "Stanford President Sets Initiative on Teaching." *Washington Post,* March 3, 1991, p. A12.
104. Ibid.
105. Dennis Hayes and Robin Wynyard. "Introduction." In Dennis Hayes and Robin Wynyard, eds., *The McDonaldization of Higher Education.* Westport, CT: Bergin & Garvey, 2002, p. 11.
106. An example of hundreds of DRGs is DRG 236, "Fractures of Hip and Pelvis." A set amount is reimbursed by Medicare for medical procedures included under that heading and all other DRGs.
107. Dan Colburn. "Unionizing Doctors: Physicians Begin Banding Together to Fight for Autonomy and Control Over Medical Care." *Washington Post/Health,* June 19, 1985, p. 7.
108. Sports are not alone in this; the political parties have shortened and streamlined their conventions to accommodate the needs and demands of television.
109. Allen Guttman. *From Ritual to Record: The Nature of Modern Sports.* New York: Cambridge University Press, 1978, p. 47.
110. Ibid., p. 51.
111. For those unfamiliar with baseball, a designated hitter is one of a team's starting players and takes a regular turn at bat throughout a game. A pinch hitter comes in during a game and bats for one of the players in the game. Pinch hitters almost always get only that one turn at bat during the game.
112. However, specialization in baseball has more than compensated for this, and it is undoubtedly the case that people now see more rather than less use of relief pitchers. Indeed, there are now very specialized relief roles—the "long reliever" who comes in early in the game, the "closer" who finishes off a game in which his team is ahead, and relievers who specialize in getting out left- or right-handed batters.
113. http://archives.cbc.ca/IDD-1-41-1727/sports/extreme_sports/, retrieved March 15, 2007
114. Carl Schoettler. "Examining the Pull of the Poll." *Sun* (Baltimore), October 11, 1998, pp. 13Fff.
115. Kathleen Jamieson. *Eloquence in an Electronic Age: The Transformation of Political Speechmaking.* New York: Oxford University Press, 1988, p. 11.
116. Ibid.; see also Marvin Kalb. "TV, Election Spoiler." *New York Times,* November 28, 1988, p. A25.
117. www.youtube.com/profile?user=BarackObamadotcom#g/u
118. Ester Reiter. *Making Fast Food.* Montreal: McGill-Queen's University Press, 1991, p. 85.
119. Jill Lawrence. "80 Pizzas Per Hour." *Washington Post,* June 9, 1996, pp. W07ff.
120. Stan Luxenberg. *Roadside Empires: How the Chains Franchised America.* New York: Viking, 1985, pp. 73–74.
121. www.washingtonpost.com/lifestyle/magazine/whos-lovin-it/20
122. Ibid.
123. Stan Luxenberg. *Roadside Empires: How the Chains Franchised America.* New York: Viking, 1985, p. 80.
124. Ibid., pp. 84–85.
125. Robin Leidner. *Fast Food, Fast Talk: Service Work and the Routinization of Everyday Life.* Berkeley: University of California Press, 1993, p. 60.
126. Stuart Flexner. *I Hear America Talking.* New York: Simon & Schuster, 1976, p. 142.

127. Frederick W. Taylor. *The Principles of Scientific Management.* New York: Harper & Row, 1947, p. 42.
128. Ibid., p. 138.
129. Mark Dowie. "Pinto Madness." *Mother Jones,* September/October 1977, pp. 24ff.

Chapter 4

1. W. Baldamus. "Tedium and Traction in Industrial Work." In David Weir, ed., *Men and Work in Modern Britain.* London: Fontana, 1973, pp. 78–84.
2. www.bestwestern.com/newsroom/factsheet_countrydetail.asp
3. InterContinental Hotels Group website: www.ihgplc.com/files/pdf/ factsheets/ihg_at_a_ glance.pdf
4. www.wyndhamworldwide.com/about/hojo.cfm?cid=HJWYNWW
5. Entrepreneur website: www.entrepreneur.com
6. Robin Leidner. *Fast Food, Fast Talk: Service Work and the Routinization of Everyday Life.* Berkeley: University of California Press, 1993, pp. 45–47, 54.
7. Cited in Ibid., p. 82.
8. Margaret King. "McDonald's and the New American Landscape." *USA TODAY,* January 1980, p. 46.
9. Duke Helfand. "A Super-Sized Way to Worship; California Has More Megachurches Than Any Other State, With the Majority in Suburbs Between Los Angeles and San Diego." *Los Angeles Times,* October 11, 2009, Part A, p. 41.
10. http://hirr.hartsem.edu/cgi-bin/mega/db.pl?db=default&uid=default&view_ records=1&ID=*&sb=3&so=descend
11. Jacqueline L. Salmon and Hamil R. Harris. "Reaching Out With the Word—And Technology." *Washington Post,* February 4, 2007, pp. A1, A8.
12. From the song "Little Boxes." Words and music by Malvina Reynolds. Copyright 1962, Schroder Music Co. (ASCAP). Renewed 1990. Used by permission. All rights reserved.
13. Marcus Palliser. "For Suburbia Read Fantasia: Disney Has Created the American Dream Town in Sunny Florida." *Daily Telegraph,* November 27, 1996, pp. 31ff.
14. http://haircuttery.com/home/more/aboutus/about_us.dot
15. www.washingtonpost.com/lifestyle/magazine/whos-lovin-it/20
16. Robin Leidner. *Fast Food, Fast Talk: Service Work and the Routinization of Everyday Life.* Berkeley: University of California Press, 1993, p. 82.
17. Ibid.
18. Ibid., pp. 220, 230.
19. Ibid.
20. I will have more to say about this aspect of McDonaldization in Chapter 5.
21. Robin Leidner. *Fast Food, Fast Talk: Service Work and the Routinization of Everyday Life.* Berkeley: University of California Press, 1993, pp. 107, 108.
22. Elspeth Probyn. "McIdentities: Food and the Familial Citizen." *Theory, Culture and Society* 15 (1998): 155–173.
23. Robin Leidner. *Fast Food, Fast Talk: Service Work and the Routinization of Everyday Life.* Berkeley: University of California Press, 1993, p. 10.
24. Leidner reports that employees are encouraged to vary the process in order to reduce the customers' feelings of depersonalization. But at the franchise in which she worked, limits were placed on even this.
25. Robin Leidner. *Fast Food, Fast Talk: Service Work and the Routinization of Everyday Life.* Berkeley: University of California Press, 1993, p. 25.
26. Ibid.
27. Harrison M. Trice and Janice M. Beyer. *The Cultures of Work Organizations.* Englewood Cliffs, NJ: Prentice Hall, 1993.

28. Mary-Angie Salva-Ramirez. "McDonald's: A Prime Example of Corporate Culture." *Public Relations Quarterly,* December 22, 1995, pp. 30ff.
29. www.aboutmcdonalds/mcd/careers/hamburger_university.html; Dick Schaaf. "Inside Hamburger University." *Training,* December 1994, pp. 18–24.
30. Robin Leidner. *Fast Food, Fast Talk: Service Work and the Routinization of Everyday Life.* Berkeley: University of California Press, 1993, p. 58.
31. http://wdw.disneycareers.com/en/working-here/the-disney-look
32. Lynn Darling. "On the Inside at Parks à la Disney." *Washington Post,* August 28, 1978, p. A10.
33. Ibid.
34. Robin Leidner. *Fast Food, Fast Talk: Service Work and the Routinization of Everyday Life.* Berkeley: University of California Press, 1993, p. 58.
35. Henry Mitchell. "Wonder Bread, Any Way You Slice It." *Washington Post,* March 22, 1991, p. F2.
36. William Serrin. "Let Them Eat Junk." *Saturday Review,* February 2, 1980, p. 18.
37. Matthew Gilbert. "In McMovieworld, Franchises Taste Sweetest." *Commercial Appeal* (Memphis), May 30, 1997, pp. E10ff.
38. John Powers. "Tales of Hoffman." *Washington Post,* Sunday Arts, March 5, 1995, p. G6.
39. Matthew Gilbert. "TV's Cookie-Cutter Comedies." *Boston Globe,* October 19, 1997, pp. N1ff.
40. Ibid.
41. Ibid.
42. Similarly, Busch Gardens offers European attractions, such as a German-style beer hall, without having its clientele leave the predictable confines of the United States and the even more predictable surroundings of the modern amusement park.
43. At the opening of the Istanbul Hilton, Conrad Hilton said, "Each of our hotels . . . is a 'little America.'" This quotation is from Daniel J. Boorstin. *The Image: A Guide to Pseudo-Events in America.* New York: Harper Colophon, 1961, p. 98.
44. John Urry. *The Tourist Gaze: Leisure and Travel in Contemporary Societies.* London: Sage, 1990.
45. Andrew Beyer. "Lukas Has the Franchise on Almighty McDollar." *Washington Post,* August 8, 1990, pp. F1, F8.
46. William Severini Kowinski. *The Malling of America: An Inside Look at the Great Consumer Paradise.* New York: William Morrow, 1985, p. 27.
47. Iver Peterson. "Urban Dangers Send Children Indoors to Play: A Chain of Commercial Playgrounds Is One Answer for Worried Parents." *New York Times,* January 1, 1995, sec. 1, p. 29.
48. Jan Vertefeuille. "Fun Factory: Kids Pay to Play at the Discovery Zone and While That's Just Fine With Many Parents, It Has Some Experts Worried." *Roanoke Times & World News,* December 8, 1994, Extra, pp. 1ff.
49. Cited in Stephen J. Fjellman. *Vinyl Leaves: Walt Disney World and America.* Boulder, CO: Westview, 1992, p. 226.
50. Dirk Johnson. "Vacationing at Campgrounds Is Now Hardly Roughing It." *New York Times,* August 28, 1986, p. B1.
51. http://koapressroom.com/companyinformation/; "CountryClub Campgrounds." *Newsweek,* September 24, 1984, p. 90.
52. Dirk Johnson. "Vacationing at Campgrounds Is Now Hardly Roughing It." *New York Times,* August 28, 1986, p. B1.
53. Kristin Downey Grimsley. "Risk of Homicide Is Higher in Retail Jobs: Half of Workplace Killings Sales-Related." *Washington Post,* July 13, 1997, pp. A14ff.
54. Richard Edwards. *Contested Terrain: The Transformation of the Workplace in the Twentieth Century.* New York: Basic Books, 1979.
55. Erik Brynjolfsson and Andrew McAfee. *Race Against the Machine: How the Digital Revolution Is Accelerating Innovation, Driving Productivity, and Irreversibly Transforming Employment and the Economy.* Kindle Books, 2011.

56. Richard Edwards. *Contested Terrain: The Transformation of the Workplace in the Twentieth Century.* New York: Basic Books, 1979.

57. Jerry Newman. *My Secret Life on the McJob: Lessons From Behind the Counter Guaranteed to Supersize Any Management Style.* New York: McGraw-Hill, 2007, p. 52.

58. http://wearemjr.com/2011/06/06/the-burger-that-ate-britain/

59. Michael Lev. "Raising Fast Food's Speed Limit." *Washington Post,* August 7, 1991, pp. D1, D4.

60. Ray Kroc. *Grinding It Out.* New York: Berkeley Medallion, 1977, pp. 131–132.

61. Eric A. Taub. "The Burger Industry Takes a Big Helping of Technology." *New York Times,* October 8, 1998, pp. 13Gff.

62. William R. Greer. "Robot Chef's New Dish: Hamburgers." *New York Times,* May 27, 1987, p. C3.

63. Ibid.

64. Michael Lev. "Taco Bell Finds Price of Success (59 cents)." *New York Times,* December 17, 1990, p. D9.

65. Calvin Sims. "Robots to Make Fast Food Chains Still Faster." *New York Times,* August 24, 1988, p. 5.

66. Chuck Murray. "Robots Roll From Plant to Kitchen." *Chicago Tribune–Business,* October 17, 1993, pp. 3ff.

67. Eric A. Taub. "The Burger Industry Takes a Big Helping of Technology." *New York Times,* October 8, 1998, pp. 13Gff.

68. www.kueducation.com/us

69. KinderCare website: www.kindercare.com/about/

70. "The McDonald's of Teaching." *Newsweek,* January 7, 1985, p. 61.

71. http://tutoring.sylvanlearning.com/sylvan_about_us.cfm; http://tutoring.sylvanlearning .com/ about-us/index.cfm

72. "The McDonald's of Teaching." *Newsweek,* January 7, 1985, p. 61.

73. William Stockton. "Computers That Think." *New York Times Magazine,* December 14, 1980, p. 48.

74. Bernard Wysocki Jr. "Follow the Recipe: Children's Hospital in San Diego Has Taken the Standardization of Medical Care to an Extreme." *Wall Street Journal,* April 22, 2003, pp. R4ff.

75. Frederick W. Taylor. *The Principles of Scientific Management.* New York: Harper & Row, 1947, p. 59.

76. Henry Ford. *My Life and Work.* Garden City, NY: Doubleday, 1922, p. 103.

77. Robin Leidner. *Fast Food, Fast Talk: Service Work and the Routinization of Everyday Life.* Berkeley: University of California Press, 1993, p. 105.

78. Virginia A. Welch. "Big Brother Flies United." *Washington Post–Outlook,* March 5, 1995, p. C5.

79. Ibid.

80. StopJunkCalls website: www.stopjunkcalls.com/convict.htm, retrieved May 2007.

81. Gary Langer. "Computers Reach Out, Respond to Human Voice." *Washington Post,* February 11, 1990, p. H3.

82. Carl H. Lavin. "Automated Planes Raising Concerns." *New York Times,* August 12, 1989, pp. 1, 6.

83. Robin Leidner. *Fast Food, Fast Talk: Service Work and the Routinization of Everyday Life.* Berkeley: University of California Press, 1993.

84. L. B. Diehl and M. Hardart. *The Automat: The History, Recipes, and Allure of Horn and Hardart's Masterpiece.* New York: Clarkson Potter, 2002.

85. Stan Luxenberg. *Roadside Empires: How the Chains Franchised America.* New York: Viking, 1985.

86. Martin Plimmer. "This Demi-Paradise: Martin Plimmer Finds Food in the Fast Lane Is Not to His Taste." *Independent* (London), January 3, 1998, p. 46.

87. Harold Gracey. "Learning the Student Role: Kindergarten as Academic Boot Camp." In Dennis Wrong and Harold Gracey, eds., *Readings in Introductory Sociology.* New York: Macmillan, 1967, pp. 243–254.

88. Charles E. Silberman. *Crisis in the Classroom: The Remaking of American Education.* New York: Random House, 1970, p. 122.

89. Ibid., p. 137.

90. Ibid., p. 125.

91. William Severini Kowinski. *The Malling of America: An Inside Look at the Great Consumer Paradise.* New York: William Morrow, 1985, p. 359.

92. Jerry Newman. *My Secret Life on the McJob: Lessons From Behind the Counter Guaranteed to Supersize Any Management Style.* New York: McGraw-Hill, 2007, p. 21. However, Newman found that there was much greater variation from store to store on personnel procedures (e.g., the hiring, training, and motivating of employees).

93. William Serrin. "Let Them Eat Junk." *Saturday Review,* February 2, 1980, p. 23.

94. AquaSol, Inc. website: www.fishfarming.com

95. Juliet Eilperin. "Farm-Fresh Fish—With a Catch; Acquaculture Boom Raises Concerns." *Washington Post,* September 20, 2009: A03; Cornelia Dean, "Rules Guiding Fish Farming in the Gulf Are Readied." *New York Times,* September 4, 2009, p. A16; Martha Duffy. "The Fish Tank on the Farm." *Time,* December 3, 1990, pp. 107–111.

96. Peter Singer. *Animal Liberation: A New Ethic for Our Treatment of Animals.* New York: Avon, 1975.

97. Ibid., pp. 96–97.

98. Ibid., pp. 105–106.

99. Ibid., p. 123.

100. Lenore Tiefer. "The Medicalization of Impotence: Normalizing Phallocentrism." *Gender & Society* 8 (1994): 363–377.

101. Cheryl Jackson. "Impotence Clinic Grows Into Chain." *Tampa Tribune—Business and Finance,* February 18, 1995, p. 1.

102. L. Mamo and J. R. Fishman. "Potency in All the Right Places: Viagra as a Technology of the Gendered Body." *Body & Society* 7 (2001): 13. As has been made clear, Viagra serves in many ways to McDonaldize sex. Its use has become a phenomenon, and a subject of concern (and some humor), not only in the United States but elsewhere in the world. In Spain, there have been robberies focusing on stealing Viagra from pharmacies. It has become a recreational drug there, demanded even by young people, which has led to enormous sales even at high retail prices ($104 for a box of eight) and to illegal sales (at discos, for example), with one pill going for as much as $80. Why this great demand in a society noted for its macho culture? According to a spokesperson for Pfizer, the maker of Viagra, it is linked to McDonaldization: "We used to have a siesta, to sleep all afternoon. . . . But now we have *become a fast-food nation* where everyone is stressed out, and this is not good for male sexual performance" (italics added). Dan Bilefsky. "Spain Says Adios Siesta and Hola Viagra." *New York Times,* February 11, 2007, p. 14.

103. Annette Baran and Reuben Pannor. *Lethal Secrets: The Shocking Consequences and Unresolved Problems of Artificial Insemination.* New York: Warner, 1989.

104. Paula Mergenbagen DeWitt. "In Pursuit of Pregnancy." *American Demographics,* May 1993, pp. 48ff.

105. Eric Adler. "The Brave New World: It's Here Now, Where In Vitro Fertilization Is Routine and Infertility Technology Pushes Back All the Old Limitations." *Kansas City Star,* October 25, 1998, pp. G1ff.

106. Clear Passage website: www.clearpassage.com/about_infertility_therapy.htm, retrieved May 2007.

107. *Drug Week,* October 24, 2008, p. 1487.

108. "No Price for Little Ones." *Financial Times,* September 28, 1998, pp. 17ff.

109. Diederika Pretorius. *Surrogate Motherhood: A Worldwide View of the Issues.* Springfield, IL: Charles C Thomas, 1994.

110. Korky Vann. "With In-Vitro Fertilization, Late-Life Motherhood Becoming More Common." *Hartford Courant,* July 7, 1997, pp. E5ff.

111. http://news.bbc.co.uk/2/hi/europe/4199839.stm, retrieved May 2007.

112. Angela Cain. "Home Test Kits Fill an Expanding Health Niche." *Times Union-Life and Leisure* (Albany, NY), February 12, 1995, p. 11.

113. Neil Bennett, ed. *Sex Selection of Children*. New York: Academic Press, 1983.
114. "Selecting Sex of Child." *South China Morning Post*, March 20, 1994, p. 15; www.havingbabies .com/gender-selection.html, retrieved May 2007.
115. www.havingbabies.com/gender-selection.html, retrieved May 2007.
116. Janet Daley. "Is Birth Ever Natural?" *The Times* (London), March 16, 1994, p. 18.
117. Matt Ridley. "A Boy or a Girl: Is It Possible to Load the Dice?" *Smithsonian* 24 (June 1993): 123.
118. Roger Gosden. *Designing Babies: The Brave New World of Reproductive Technology*. New York: W. H. Freeman, 1999, p. 243.
119. Rayna Rapp. "The Power of 'Positive' Diagnosis: Medical and Maternal Discourses on Amniocentesis." In Donna Bassin, Margaret Honey, and Meryle Mahrer Kaplan, eds., *Representations of Motherhood*. New Haven, CT: Yale University Press, 1994, pp. 204–219.
120. Aliza Kolker and B. Meredith Burke. *Prenatal Testing: A Sociological Perspective*. Westport, CT: Bergin & Garvey, 1994, p. 158.
121. Jeffrey A. Kuller and Steven A. Laifer. "Contemporary Approaches to Prenatal Diagnosis." *American Family Physician* 52 (December 1996): 2277ff.
122. Aliza Kolker and B. Meredith Burke. *Prenatal Testing: A Sociological Perspective*. Westport, CT: Bergin & Garvey, 1994; Ellen Domke and Al Podgorski. "Testing the Unborn: Genetic Test Pinpoints Defects, but Are There Risks?" *Chicago Sun-Times*, April 17, 1994, p. C5.
123. However, some parents do resist the rationalization introduced by fetal testing. See Shirley A. Hill. "Motherhood and the Obfuscation of Medical Knowledge." *Gender & Society* 8 (1994): 29–47.
124. Mike Chinoy. CNN News. February 8, 1994.
125. Joan H. Marks. "The Human Genome Project: A Challenge in Biological Technology." In Gretchen Bender and Timothy Druckery, eds., *Culture on the Brink: Ideologies of Technology*. Seattle, WA: Bay Press, 1994, pp. 99–106; R. C. Lewontin. "The Dream of the Human Genome." In Gretchen Bender and Timothy Druckery, eds., *Culture on the Brink: Ideologies of Technology*. Seattle, WA: Bay Press, 1994, pp. 107–127.
126. "Genome Research: International Consortium Completes Human Genome Project." *Genomics & Genetics Weekly*, May 9, 2003, p. 32.
127. Matt Ridley. "A Boy or a Girl: Is It Possible to Load the Dice?" *Smithsonian* 24 (June 1993): 123.
128. Jessica Mitford. *The American Way of Birth*. New York: Plume, 1993.
129. For a critique of midwifery from the perspective of rationalization, see Charles Krauthammer. "Pursuit of a Hallmark Moment Costs a Baby's Life." *Tampa Tribune*, May 27, 1996, p. 15.
130. American College of Nurse-Midwives, 2008 Report; Judy Foreman. "The Midwives' Time Has Come—Again." *Boston Globe*, November 2, 1998, pp. C1ff.
131. www.allnursingschools.com/faqs/cnm.php; retrieved March 2007.
132. Jessica Mitford. *The American Way of Birth*. New York: Plume, 1993, p. 13.
133. Catherine Kohler Riessman. "Women and Medicalization: A New Perspective." In P. Brown, ed., *Perspectives in Medical Sociology*. Prospect Heights, IL: Waveland, 1989, pp. 190–220.
134. Michelle Harrison. *A Woman in Residence*. New York: Random House, 1982, p. 91.
135. Judith Walzer Leavitt. *Brought to Bed: Childbearing in America, 1750–1950*. New York: Oxford University Press, 1986, p. 190.
136. Ibid.
137. Paula A. Treichler. "Feminism, Medicine, and the Meaning of Childbirth." In Mary Jacobus, Evelyn Fox Keller, and Sally Shuttleworth, eds., *Body Politics: Women and the Discourses of Science*. New York: Routledge, 1990, pp. 113–138.
138. Jessica Mitford. *The American Way of Birth*. New York: Plume, 1993, p. 59.
139. An episiotomy is an incision from the vagina toward the anus to enlarge the opening needed for a baby to pass.

140. Jessica Mitford. *The American Way of Birth*. New York: Plume, 1993, p. 61.
141. Ibid., p. 143.
142. Michelle Harrison. *A Woman in Residence*. New York: Random House, 1982, p. 86.
143. www.bmj.com/cgi/congtent/full/328/745/11302?ct, retrieved March 2007.
144. Michelle Harrison. *A Woman in Residence*. New York: Random House, 1982, p. 113.
145. Jeanne Guillemin. "Babies by Cesarean: Who Chooses, Who Controls?" In P. Brown, ed., *Perspectives in Medical Sociology*. Prospect Heights, IL: Waveland, 1989, pp. 549–558.
146. L. Silver and S. M. Wolfe. *Unnecessary Cesarean Sections: How to Cure a National Epidemic*. Washington, DC: Public Citizen Health Research Group, 1989.
147. Joane Kabak. "C Sections." *Newsday,* November 11, 1996, pp. B25ff.
148. www.cdc.gov/nchs/pressroom/obfacts/birth05.htm, retrieved March 2007; www.cdc.gov/nchs/data/databriefs/db35.pdf; Denise Grady. "Caesarean Births Are at a High in U.S." *New York Times,* March 23, 2010.
149. Susan Brink. "Too Posh to Push?" *U.S. News & World Report,* August 5, 2002, Health and Medicine sec., p. 42.
150. Randall S. Stafford. "Alternative Strategies for Controlling Rising Cesarean Section Rates." *JAMA* 263 (1990): 683–687.
151. Jeffrey B. Gould, Becky Davey, and Randall S. Stafford. "Socioeconomic Differences in Rates of Cesarean Sections." *New England Journal of Medicine* 321 (1989): 233–239; F. C. Barros, J. P. Vaughan, C. G. Victora, and S. R. Huttly. "Epidemic of Caesarean Sections in Brazil." *The Lancet* 338 (1991): 167–169.
152. Randall S. Stafford. "Alternative Strategies for Controlling Rising Cesarean Section Rates." *JAMA* 263 (1990): 683–687.
153. Although, more recently, insurance and hospital practices have led to more deaths in nursing homes or even at home.
154. Sherwin B. Nuland. *How We Die: Reflections on Life's Final Chapter*. New York: Knopf, 1994, p. 255; National Center for Health Statistics. *Vital Statistics of the United States, 1992–1993, Vol. 2, Mortality, Part A*. Hyattsville, MD: Public Health Service, 1995; www.nlm.nih.gov/medlineplus/news/fullstory_91474.html; www.ncbi.nlm.nih.gov/pubmed/18043014; www.nhpco.org/files/public/statistics_Research/ NHPCO_facts_and_figures.pdf
155. Derek Humphry. *Final Exit: The Practicalities of Self-Deliverance and Assisted Suicide for the Dying*, 3rd ed. New York: Delta, 2002.
156. Richard A. Knox. "Doctors Accepting of Euthanasia, Poll Finds: Many Would Aid in Suicide Were It Legal." *Boston Globe,* April 23, 1998, pp. A5ff.
157. A. Gruneir, V. Mor, S. Weitzen, R. Truchil, J. Teno, and J. Roy. "Where People Die: A Multilevel Approach to Understanding Influences on Site of Death in America." *Medical Care Research and Review* 64 (2007): 351; Katie Zezima. "Home Burials Offering an Intimate Alternative, at a Lower Cost." *New York Times,* July 21, 2009, pp. A1ff.
158. Ellen Goodman. "Kevorkian Isn't Helping 'Gentle Death.'" *Newsday,* August 4, 1992, p. 32.
159. Lance Morrow. "Time for the Ice Floe, Pop: In the Name of Rationality, Kevorkian Makes Dying—and Killing—Too Easy." *Time,* December 7, 1998, pp. 48ff.

Chapter 5

1. Negative effects other than the ones discussed here, such as racism and sexism, cannot be explained by this process. See Ester Reiter. *Making Fast Food*. Montreal: McGill-Queen's University Press, 1991, p. 145.
2. As they are to the critical theorists. See, for example, Martin Jay. *The Dialectical Imagination*. Boston: Little, Brown, 1973.
3. Julie Jargon. "McD's Service Stalls at Drive-Thru." *Crain's Chicago Business,* January 2, 2006, pp. 1ff.

4. America's Best Drive Thru—QSR Drive Thru Performance Study: www.qsrmagazine .com/reports/drive-thru_time_study/index.phtml
5. Julie Jargon. "McD's Service Stalls at Drive-Thru." *Crain's Chicago Business,* January 2, 2006, pp. 1ff.
6. Ibid.
7. Mike Comerford. "The Forbidden Kitchen Technology and Testing Help: McDonald's Grow Around the Globe." *Chicago Daily Herald-Business,* December 11, 2006, pp. 1ff.
8. http://nerdynerdnerdz.com/2011/07/mcdonalds-restaurants-launch-new-drive-thru-tablets/
9. Michael Schrage. "The Pursuit of Efficiency Can Be an Illusion." *Washington Post,* March 20, 1992, p. F3.
10. Richard Cohen. "Take a Message—Please!" *Washington Post Magazine,* August 5, 1990, p. 5.
11. Peter Perl. "Fast Is Beautiful." *Washington Post Magazine,* May 24, 1992, p. 26.
12. Mark Bittman. "Is Junk Food Really Cheaper?" *New York Times Sunday Review,* September 25, 2011, pp. 1, 7.
13. http://opinionator.blogs.nytimes.com/2011/02/22/how-to-make-oatmeal-wrong/?emc=etal
14. Melanie Warner. "Salads or No: Cheap Burgers Revive McDonald's." *New York Times,* April 19, 2006, pp. A1ff.
15. Bob Garfield. "How I Spent (and Spent and Spent) My Disney Vacation." *Washington Post/ Outlook,* July 7, 1991, p. B5.
16. Ibid.
17. Ester Reiter. *Making Fast Food.* Montreal: McGill-Queen's University Press, 1991, p. 95.
18. Jill Smolowe. "Read This!!!!" *Time,* November 26, 1990, pp. 62ff.
19. Michael Schrage. "Personalized Publishing: Confusing Information With Intimacy." *Washington Post,* November 23, 1990, p. B13.
20. Mark A. Schneider. *Culture and Enchantment.* Chicago: University of Chicago Press, 1993, p. ix. Weber derived this notion from Friedrich Schiller.
21. Hans Gerth and C. Wright Mills. "Introduction." In Hans Gerth and C. Wright Mills, eds., *From Max Weber.* New York: Oxford University Press, 1958, p. 51.
22. Mark A. Schneider. *Culture and Disenchantment.* Chicago: University of Chicago Press, 1993, p. ix.
23. Virginia Stagg Elliott. "Fast-Food Sellers Under Fire for Helping Supersize People." *American Medical News,* April 21, 2003. See www.ama-assn.org/amednews/2003/04/21/hlsc0421.htm, retrieved May 2007.
24. Mark Bittman. "Is Junk Food Really Cheaper?" *New York Times Sunday Review,* September 25, 2011, p. 7.
25. Jeremy Laurance. "Slow Killer on the March." *Toronto Star,* March 4, 2006, p. D12.
26. David A. Alfer and Karen Eny. "The Relationship Between the Supply of Fast-Food Chains and Cardiovascular Outcomes." *Canadian Journal of Public Health* 96 (May 2001): 173ff; Sharon Kirkey. "Nutrition: New Study Links Fast-Food Spots, Death Rates." *National Post,* May 12, 2005, p. A15.
27. Maryellen Spencer. "Can Mama Mac Get Them to Eat Spinach?" In Marshall Fishwick, ed., *Ronald Revisited: The World of Ronald McDonald.* Bowling Green, OH: Bowling Green University Press, 1983, pp. 85–93.
28. Donald J. Hernandez and Evan Charney, eds. *From Generation to Generation: The Health and Well-Being of Children in Immigrant Families.* Washington, DC: National Academy Press, 1998.
29. Rachel Abramowitz. "Disney Loses Its Appetite for Happy Meal Tie-Ins." *Los Angeles Times,* May 8, 2006, pp. A1ff. However, McDonald's restaurants will remain open in Disney parks, and there is the possibility of adult cross-promotions in the future.
30. Patty Lanoue Stearns. "Double-Sized Fast Foods Means Double the Trouble." *Pittsburgh Post-Gazette,* October 10, 1996, p. B6.
31. http://nutrition.mcdonalds.com/nutritionexchange/nutritionfacts.pdf
32. Ibid; http://burgerking.co.uk/nutrition?producttypeid=178&productid=4

33. http://opinionator.blogs.nytimes.com/2011/02/22/how-to-make-oatmeal-wrong/?emc=etal
34. Ibid.
35. Ibid.
36. Ibid.
37. "At 42, Ronald McDonald Reborn as Fitness Fanatic." *Charleston Daily Mail,* June 10, 2005, p. 1a.
38. Regina Schrambling. "The Curse of Culinary Convenience." *New York Times,* September 10, 1991, p. A19.
39. Ibid.
40. "*E. coli* Outbreak Forces Closure of Meat Plant." *Independent* (London), August 22, 1997, p. 12.
41. Erin Allday. "Technology, Eating Habits Help to Spread *E. coli.*" *San Francisco Chronicle,* September 23, 2006, p. A9.
42. http://abcnews.go.com/GMA/HealthyLiving/coli-spinach-salad-safe/story? id=9034833
43. www.worldwatch.org/node/5443
44. Arthur Beesley. "China's Diet Revolution Threatens the Environment." *Irish Times,* December 28, 2006, p. 8.
45. Bill Bell Jr. "Environmental Groups Seeking Moratorium on New or Expanded 'Animal Factories.'" *St. Louis Post-Dispatch,* December 4, 1998, p. C8.
46. Tim O'Brien. "Farming: Poison Pens." *Guardian* (London), April 29, 1998, p. 4.
47. Olivia Wu. "Raising Questions: Environmentalists Voice Concerns Over Booming Aquaculture Industry." *Chicago Tribune,* September 9, 1998, pp. 7Aff; Colin Woodard. "Fish Farms Get Fried for Fouling." *Christian Science Monitor,* September 9, 1998, pp. 1ff.
48. Timothy Egan. "In Land of French Fry, Study Finds Problems." *New York Times,* February 7, 1994, p. A10.
49. Max Boas and Steve Chain. *Big Mac: The Unauthorized Story of McDonald's.* New York: E. P. Dutton, 1976.
50. Al Gore. *An Inconvenient Truth: The Planetary Emergency of Global Warming and What We Can Do About It.* New York: Rodale Press, 2006.
51. In many areas, there has been a simultaneous increase in reasonably authentic ethnic restaurants.
52. "The Grand Illusion." *The Economist,* June 5, 1999, pp. 2–18.
53. Ellen Goodman. "Fast-Forwarding Through Fall." *Washington Post,* October 5, 1991, p. A19. There is another irrationality here. Those who buy things through catalogs find that their deliveries are often late or they never arrive at all. Said the president of the Better Business Bureau of Metropolitan New York, "With mail order, the biggest problem is delivery and delay in delivery." See Leonard Sloane. "Buying by Catalogue Is Easy: Timely Delivery May Not Be." *New York Times,* April 25, 1992, p. 50.
54. George Ritzer. *The McDonaldization Thesis.* London: Sage, 1998, pp. 59–70; Jerry Newman. *My Secret Life on the McJob: Lessons From Behind the Counter Guaranteed to Supersize Any Management Style.* New York: McGraw-Hill, 2007; Rachel Osterman. "Flipping an Image: McDonald's Battles the Stigma That Fast Food Work Is a Dead-end Job." *Sacramento Bee,* October 14, 2005, p. D1; Roland Gribben. "McDonald's Dispels Negative Image." *Daily Telegraph,* June 8, 2006, p. 1.
55. www.washingtonpost.com/lifestyle/magazine/who's-lovin-it/20
56. Ester Reiter. *Making Fast Food.* Montreal: McGill-Queen's University Press, 1991, pp. 150, 167.
57. Leidner disagrees with this, arguing that McDonald's "workers expressed relatively little dissatisfaction with the extreme routinization." See Robin Leidner. *Fast Food, Fast Talk: Service Work and the Routinization of Everyday Life.* Berkeley: University of California Press, 1993, p. 134. One could ask, however, whether this indicates a McDonaldizing society in which people, accustomed to the process, simply accept it as an inevitable part of their work.
58. Another estimate puts such turnover between 200% and 250%; see Jerry Newman. *My Secret Life on the McJob: Lessons From Behind the Counter Guaranteed to Supersize Any Management Style.* New York: McGraw-Hill, 2007, p. 167.

59. Henry Ford. *My Life and Work.* Garden City, NY: Doubleday Page, 1922, pp. 105, 106.
60. Studs Terkel. *Working.* New York: Pantheon, 1974, p. 159.
61. Barbara Garson. *All the Livelong Day.* Harmondsworth, UK: Penguin, 1977, p. 88.
62. Studs Terkel. *Working.* New York: Pantheon, 1974, p. 175.
63. For a review of the literature on this issue, see George Ritzer and David Walczak. *Working: Conflict and Change,* 3rd ed. Englewood Cliffs, NJ: Prentice Hall, 1986, pp. 328–372.
64. Eric Schlosser. *Fast Food Nation: The Dark Side of the All-American Meal.* Boston: Houghton Mifflin, 2001.
65. www.washingtonpost.com/lifestyle/magazine/who's-lovin-it/20
66. Robin Leidner. *Fast Food, Fast Talk: Service Work and the Routinization of Everyday Life.* Berkeley: University of California Press, 1993, p. 30.
67. Bob Garfield. "How I Spent (and Spent and Spent) My Disney Vacation." *Washington Post/Outlook,* July 7, 1991, p. 5.
68. Ray Oldenburg. *The Great Good Place.* New York: Paragon, 1987.
69. One exception to the general rule that diners do not linger is the tendency for retirees to use McDonald's as a social center, especially over breakfast or coffee. Some McDonald's restaurants even allow seniors to conduct bingo games.
70. William R. Mattox Jr. "The Decline of Dinnertime." *Ottawa Citizen,* April 30, 1997, p. A14.
71. Nicholas von Hoffman. "The Fast-Disappearing Family Meal." *Washington Post,* November 23, 1978, p. C4.
72. www.google.com/hostednews/ap/article/ALeqM5g16Kt_aVGSPXiHJ_EmppGF4YeaKgD9BUBH4O0
73. Margaret Visser. "A Meditation on the Microwave." *Psychology Today,* December 1989, p. 42.
74. Ibid., pp. 38ff.
75. "The Microwave Cooks Up a New Way of Life." *Wall Street Journal,* September 19, 1989, p. B1.
76. Margaret Visser. "A Meditation on the Microwave." *Psychology Today,* December 1989, p. 40.
77. Ibid., p. 42.
78. Peggy Gisler and Marge Eberts. "Reader Disagrees With Advice for Mom Too Tired to Read." *Star Tribune* (Minneapolis), July 3, 1995, p. 3E.
79. Tamar Levin. "Online Enterprises Gain Foothold as Path to a College Degree." *New York Times,* August 25, 2011, pp. A1, A18.
80. William H. Honan. "Professors Battling Television Technology." *New York Times,* April 4, 1995, p. D24.
81. Tamar Levin. "Online Enterprises Gain Foothold as Path to a College Degree." *New York Times,* August 25, 2011, p. A18.
82. Doug Mann. "Will You Have Fries With Your Metaphysics? The McDonaldization of Higher Learning May Make People Feel Good, but It Is Death to Education." *London Free Press* (Ontario), February 19, 2005, pp. F1ff; Dennis Hayes. "Diploma? Is That With Fries?" *The Times Educational Supplement,* June 10, 2005, p. 6.
83. www.uapd.com
84. Kris Hundley. "The Inpatient Physician." *St. Petersburg Times,* July 26, 1998, pp. 1Hff.
85. Sherwin B. Nuland. *How We Die: Reflections on Life's Final Chapter.* New York: Knopf, 1994, p. 149.
86. Philippe Aries. *The Hour of Our Death.* New York: Knopf, 1981.
87. Sherwin B. Nuland. *How We Die: Reflections on Life's Final Chapter.* New York: Knopf, 1994, p. xv.
88. Jean Baudrillard. *Symbolic Exchange and Death.* London: Sage, 1976/1993, p. 180.
89. Nancy Gibbs. "Rx for Death." *Time,* May 31, 1993, p. 34.
90. Sherwin B. Nuland. *How We Die: Reflections on Life's Final Chapter.* New York: Knopf, 1994, p. 254.

Chapter 6

1. Vic Sussman. "The Machine We Love to Hate." *Washington Post Magazine*, June 14, 1987, p. 33.
2. Ibid., p. 33.
3. Tanya Wenman Steel. "Have Time to Bake? What a Luxury!" *New York Times*, February 8, 1995, p. C4.
4. Weber, cited in Hans Gerth and C. Wright Mills, eds. *From Max Weber.* New York: Oxford University Press, 1958, p. 128.
5. The threefold typology presented here is not exhaustive. McDonaldized systems can also be seen as sets of "monkey bars." From this perspective, the iron cage is nothing more than a playground apparatus that can become anything the people involved with it want it to be. Thus, people can make it a velvet, rubber, or iron cage, if they so desire. While there is merit to this view, it probably overestimates the power of human beings. Cages, whether they are velvet, rubber, or iron, are structures, and therefore they (and those who support them) are often resistant to efforts to modify them. See Jay Klagge. "Approaches to the Iron Cage: Reconstructing the Bars of Weber's Metaphor." *Administration & Society* 29 (1997): 63–77.
6. Andrew Malcolm. "Bagging Old Rules to Keep a Food Co-op Viable." *New York Times*, November 8, 1991, p. B7.
7. Other examples include St. Mary's College in Maryland and Evergreen State College in Washington State.
8. www.hampshire.edu/discover/433.edu
9. June R. Herold. "B&Bs Offer Travelers Break From McBed, McBreakfast." *Business First-Columbus*, May 15, 1991, col. 1, p. 1.
10. Betsy Wade. "B&B Book Boom." *Chicago Tribune*, July 28, 1991, pp. C16ff.
11. Paul Avery. "Mixed Success for Bed-Breakfast Idea." *New York Times*, July 28, 1991, pp. 12NJ, 8.
12. Eric N. Berg. "The New Bed and Breakfast." *New York Times*, October 15, 1989, pp. 5ff.
13. Harvey Elliott. "All Mod Cons and Trouser Presses 'Ruining B&Bs.'" *Times* (London), April 3, 1996.
14. Mark Sawyer. "Avoiding Fast Foods Is Key to Fight Against Artery-Clogging Fat." *Today's Health*, June 16, 2006.
15. www.reuters.com/article/2008/07/02/newyork-transfat-idUKGR122462520080702
16. www.dietsinreview.com/diet_column/01/trans-fat-ban-starts-in-california/
17. Julie Deardorff. "If McDonald's Is Serious: Menu Needs a Makeover." *Chicago Tribune*, June 16, 2006.
18. Bruce Horovitz. "McDonald's Tinkers With Beloved Fries." *USA TODAY*, June 30, 2007.
19. John Vidal. *McLibel: Burger Culture on Trial.* New York: New Press, 1997.
20. McSpotlight website: www.mcspotlight.org
21. Ibid.
22. Danny Penman. "Judgment Day for McDonald's." *Independent* (London), June 19, 1997, pp. 20ff.
23. McSpotlight website: www.mcspotlight.org
24. Ibid.
25. Jacqueline L. Salmon. "McDonald's, Employees Reach Pact: Strike Ends." *Washington Post*, October 23, 1998, p. C3.
26. Ester Reiter. *Making Fast Food.* Montreal: McGill-Queen's University Press, 1991, pp. 70ff.
27. Workers Online website: http://workers.labor.net.au/latest, retrieved May 2007
28. David Barboza. "McDonald's in China Agrees to Unions." *New York Times*, April 10, 2007, p. C3.

29. James Brooke. "Two McDonald's in Darien Do Their Hiring in Bronx." *New York Times,* July 13, 1985, sec. 1, p. 24; Michael Winerip. "Finding a Sense of McMission in McNuggets." *New York Times,* August 23, 1988, sec. 2, p. 1; "McDonald's Seeks Retirees to Fill Void." *New York Times,* December 20, 1987, sec. 1, p. 54; Jennifer Kingson. "Golden Years Spent Under Golden Arches." *New York Times,* March 6, 1988, sec. 4, p. 26.

30. Glenn Collins. "Wanted: Child-Care Workers, Age 55 and Up." *New York Times,* December 15, 1987, sec. 1, p. 1.

31. Phyllis C. Richman. "Savoring Lunch in the Slow Lane." *Washington Post,* November 22, 1998, pp. M1ff.

32. Ibid.

33. *Slow,* July–September 1998, np.

34. www.slowfood.com/international/1/about-us?-session=query_session:AD4F3512163e72CF3 5oo2AC0746A; Corby Kummer. *The Pleasures of Slow Food.* San Francisco: Chronicle Books, 2002, p. 26.

35. Ibid., p. 23.

36. Ibid.

37. Ibid., p. 25.

38. Ibid.

39. Retrieved May 2007 from www.terramadre2004.org/terramadre/eng/comunita.lasso

40. Ibid.

41. ABC News website: http://abcnews.go.com/International/story?id=83085&page=1

42. Ibid.

43. Janet Dickinson and Les Lumsdon. *Slow Travel and Tourism.* London: Earthscan, 2010.

44. Sprawl-Busters website: www.sprawl-busters.com

45. www.sprawl-busters.com/victoryz.html

46. See Al Norman. *Slam-Dunking Wal-Mart: How You Can Stop Superstore Sprawl in Your Hometown.* Saint Johnsbury, VT: Raphael Marketing, 1999.

47. Jim Merkel. "50 Demonstrate Against McDonald's Restaurant Plan." *St. Louis Post-Dispatch,* April 14, 2006, p. A01; Hawke Fracassa. "Sterling Hts. Stops Burger King." *Detroit News,* August 14, 1998, p. C5.

48. Isabel Wilkerson. "Midwest Village; Slow-Paced, Fights Plan for Fast-Food Outlet." *New York Times,* July 19, 1987, pp. 1, 16.

49. Mary Davis Suro. "Romans Protest McDonald's." *New York Times,* May 5, 1986, p. C20.

50. Jane Perlez. "A McDonald's? Not in Their Medieval Square." *New York Times,* May 23, 1994, p. A4.

51. Dominic Kennedy. "Welcome to Burger-Free Heaven." *Times* (London), January 3, 1998.

52. Ibid.

53. In one notable exception, the entire state of Vermont had, at least until the mid-1990s, been kept free of Wal-Marts. See Paul Gruchow. "Unchaining America: Communities Are Finding Ways to Keep Independent Entrepreneurs in Business." *Utne Reader,* January-February 1995, pp. 17–18. Now, however, and consistent with the idea of increasing McDonaldization, there are a number of Wal-Marts in that state.

54. Peter Pae. "Retail Giant Rattles the Shops on Main Street." *Washington Post,* February 12, 1995, p. B3.

55. Peter Kilborn. "When Wal-Mart Pulls Out, What's Left?" *New York Times/Business,* March 5, 1995, pp. 1, 6.

56. "Eating Out Is In, and the Chains Add Variety to Lure New Diners." *Time,* August 26, 1985, pp. 60–61.

57. Anthony Ramirez. "In the Orchid Room . . . Big Macs." *New York Times,* October 30, 1990, pp. DI, D5.

58. Jane Perlez. "A McDonald's? Not in Their Medieval Square." *New York Times,* May 23, 1994, p. A4.

59. Kate Connolly. "McCafe in Vienna? Grounds for War." *Observer,* August 30, 1998, p. 19.

60. John Holusha. "McDonald's Expected to Drop Plastic Burger Box." *Washington Post,* November 1, 1990, pp. A1, D19; John Holusha. "Packaging and Public Image: McDonald's Fills a Big Order." *New York Times,* November 2, 1990, pp. Al, D5.

61. "Michigan McDonald's: McNews You Can Use." PR Newswire, June 20, 2003.
62. Warren Brown. "Hardee's to Introduce Recycled Plastic in Area." *Washington Post*, March 22, 1991, pp. B1, B3.
63. www.aboutmcdonalds.com/etc/medialib/csr/docs.Par.8633.File.dat/MCD046_BestOfGreen09_sansemails.pdf
64. Ron Alexander. "Big Mac With Chopin, Please." *New York Times*, August 12, 1990, p. 42.
65. See George Ritzer and David Walczak. *Working: Conflict and Change*, 3rd ed. Englewood Cliffs, NJ: Prentice-Hall, 1986, pp. 345–347.
66. Jerry Newman. *My Secret Life on the McJob: Lessons From Behind the Counter Guaranteed to Supersize Any Management Style*. New York: McGraw-Hill, 2007, pp. 94–95.
67. Melena Ryzik. "A New Wave Now Knits for the Rebellion of It." *New York Times*, January 27, 2007, p. A18.
68. Ibid.
69. Thomas J. Peters and Robert H. Waterman. *In Search of Excellence: Lessons From America's Best-Run Companies*. New York: Harper & Row, 1982.
70. Ibid., p. 201.
71. Robert Nelson. "Between the Lines: Changes in Industry Will Have Big Impact on What We Read and Where We Buy Our Books." *San Francisco Chronicle*, May 5, 1998, pp. E1ff.
72. Ibid.
73. Jim Dwyer. "Evicted for Manhattan Starbucks No. 188, but Defying the Coffee Octopus." *New York Times*, September 14, 2011, p. A26.
74. For a similar effort, see Neil Postman. *Technopoly*. New York: Knopf, 1992, pp. 183ff.
75. www.in-n-out.com/history.asp
76. Peter Perl. "Fast Is Beautiful." *Washington Post Magazine*, May 24, 1992, pp. 10ff; Allen Shelton. "Writing McDonald's, Eating the Past: McDonald's as a Postmodern Space." Unpublished manuscript, p. 47; Eileen Schulte. "Breakfast Club Marks Member's 99th Birthday." *St. Petersburg Times*, November 22, 1998, pp. 11ff.
77. Regina Schrambling. "The Curse of Culinary Convenience." *New York Times*, September 10, 1991, p. A19.
78. All quotations in this paragraph are from Regina Schrambling. "The Curse of Culinary Convenience." *New York Times*, September 10, 1991, p. A19.
79. Warren Leary. "Researchers Halt Ripening of Tomato." *New York Times,* October 19, 1991, p. 7.
80. John Tierney. "A Patented Berry Has Sellers Licking Their Lips." *New York Times*, October 14, 1991, p. A8.
81. James Hamilton. "Fast Food Chains Playing Pie' Piper With Mr Men and Pokemon Freebies." *The Sunday Herald* (Scotland), August 19, 2001, p. 6.
82. Eric Schmuckler. "Two Action Figures to Go, Hold the Burger." *Brandweek*, April 1, 1996, pp. 38ff.
83. Chris Rojek. *Ways of Escape: Modern Transformations in Leisure and Travel*. London: Routledge, 1993.
84. Ibid., p. 188.
85. Stanley Cohen and Laurie Taylor. *Escape Attempts: The Theory and Practice of Resistance to Everyday Life,* 2nd ed. London: Routledge, 1992.
86. James Miller. *The Passion of Michel Foucault*. New York: Anchor, 1993.
87. Stanley Cohen and Laurie Taylor. *Escape Attempts: The Theory and Practice of Resistance to Everyday Life,* 2nd ed. London: Routledge, 1992, p. 197.
88. Roger Cohen. "Faux Pas by McDonald's in Europe." *New York Times*, February 18, 1992, p. D1.
89. Two quotes from Sharon Waxman. "Paris's Sex Change Operation." *Washington Post*, March 2, 1992, p. B1.
90. "Do Not Go Gentle Into That Good Night" (excerpt). By Dylan Thomas, from *The Poems of Dylan Thomas,* copyright © 1952 by Dylan Thomas. Reprinted by permission of New Directions Publishing Corp.

Chapter 7

1. George Ritzer. *McDonaldization: The Reader,* 3rd ed. Thousand Oaks, CA: Sage, 2010.
2. Elif Izberk-Bilgin and Aaron Ahuvia. "Ebayization." In George Ritzer, ed., *McDonaldization: The Reader,* 3rd ed. Thousand Oaks, CA: Sage, 2010.
3. The others are the growing interconnectedness of the world, increasing liberalization leading to a borderless world, growing polarization leading to increasing differences between the rich and poor, Americanization, creolization leading to increasingly hybrid societies, transnationalization and the creation of transnational spaces, and balkanization leading to more clashes between different societies and civilizations. See Darren O'Byrne and Alexander Hensby. *Theorizing Global Studies.* New York: Palgrave Macmillan, 2011.
4. George Ritzer. *Globalization: A Basic Text.* Malden, MA: Wiley-Blackwell, 2010, p. 2.
5. George Ritzer. *The Globalization of Nothing 2.* Thousand Oaks, Sage, 2007.
6. George Ritzer. *The Globalization of Nothing.* Thousand Oaks, CA: Sage, 2004, p. 3. Although I use the term *generally* in this quotation, there are some forms of nothing that are locally conceived and/or controlled. Also, by *centrally* I mean, for example, the headquarters of a multinational corporation or a national government.
7. Control, as we have seen, is also a basic dimension of McDonaldization.
8. George Ritzer. *The Globalization of Nothing.* Thousand Oaks, CA: Sage, 2004, p. 7. As in the case of the caveat about the definition of nothing in regards to the term *generally,* some forms of something are centrally conceived and/or controlled.
9. For a critique of dichotomous thinking, see Elisabeth Mudimbe-Boyi, ed. *Beyond Dichotomies: Histories, Identities, Cultures, and the Challenge of Globalization.* Albany: State University of New York Press, 2002.
10. Roland Robertson. "Globalisation or Glocalisation?" *Journal of International Communication* 1(1994): 33–52.
11. For another, see Peter Berger and Samuel Huntington, eds. *Many Globalizations: Cultural Diversity in the Contemporary World.* Oxford, UK: Oxford University Press, 2002.
12. Roland Robertson. "Globalization Theory 2000+: Major Problematics." In George Ritzer and Barry Smart, eds., *Handbook of Social Theory.* London: Sage, 2001, pp. 458–471. Glocalization is at the heart of Robertson's own approach, but it is central to that of many others. The most notable is Appadurai's view that the "new global cultural economy has to be seen as a complex, overlapping, disjunctive order" (see Arjun Appadurai. *Modernity at Large: Cultural Dimensions of Globalization.* Minneapolis: University of Minnesota Press, 1996, p. 32). While John Tomlinson uses other terms, he sees glocalization as "friendly" to his own orientation (see John Tomlinson. *Globalization and Culture.* Chicago: University of Chicago Press, 1999).
13. I feel apologetic about adding yet another neologism, especially such an ungainly one, to a field already rife with jargon. However, the existence and popularity of the concept of glocalization requires the creation of the parallel notion of grobalization to emphasize what the former concept ignores or downplays.
14. I am combining a number of different entities under this heading (nations, corporations, a wide range of organizations, and so on), but it should be clear that there are profound differences among them, including the degree to which, and the ways in which, they seek to grobalize.
15. Gucci bags are nothing, as that concept is defined here, but they are certainly expensive. Grobalization as a process can also be seen in markets dealing with higher end, more expensive products as well. For example, Gucci bags, Benetton sweaters, and Prada shoes are certainly nothing, as that concept is defined here, and are also certainly considered expensive by most. Thus, the grobalization of nothing is not limited by cost considerations, although more affordable products do have more of an elective affinity with grobalization than do less affordable ones.

16. Janet Adamy. "Starbucks Bets on China's New Social Mobility." Associated Press Financial Wire, November 29, 2006.
17. Ibid.
18. Bruce Horovitz. "Starbucks Aims Beyond Lattes to Extend Brand." *USA TODAY,* May 19, 2006.
19. Janet Adamy. "Starbucks Bets on China's New Social Mobility." Associated Press Financial Wire, November 29, 2006.
20. Marc Lacey. "In Legendary Birthplace of Coffee, an Un-Starbucks." *International Herald Tribune,* July 22, 2005, p. 2.
21. "The Starbucks Index: Burgers or Beans?" *The Economist,* January 15, 2004.
22. Jacqueline L. Salmon and Hamil R. Harris. "Reaching Out With the Word—And Technology." *Washington Post,* February 4, 2007, p. A8.
23. Joseph A. Michelli. *The Starbucks Experience: 5 Principles for Turning Ordinary Into Extraordinary.* New York: McGraw-Hill, 2007.
24. Howard Schultz. *Pour Your Heart Into It: How Starbucks Built a Company One Cup at a Time.* New York: Hyperion, 1999, p. 120. www.cbsnews.com/stories/2006/04/21/60minutes/printable1532246.shtml, retrieved May 2007.
25. Ibid.
26. Randall Stross. "What Starbucks Can Learn From the Movie Palace." *New York Times—Business,* March 4, 2007, p. 3.
27. Ibid.
28. Monica Soto Ouichi. "Opportunity Brewing for Starbucks in China; Starbucks in China—Q & A With Chairman Howard Schultz." *Seattle Times,* October 9, 2005, pp. D1ff.
29. John Simmons. *My Sister's a Barista: How They Made Starbucks a Home Away From Home.* London: Cyon Books, 2005, p. 93.
30. Ibid.
31. Said newsman Scott Pelley, "It's not just coffee anymore. Starbucks is theater." *60 Minutes,* April 23, 2006.
32. L. Forlano, L. "WiFi Geographies: When Code Meets Place." *Information Society* 25 (2009): 344–352.
33. Dave Simanoff. "A Perfect Blend." *Tampa Tribune-Business,* October 29, 2006, pp. 1ff.
34. Howard Schultz. *Pour Your Heart Into It: How Starbucks Built a Company One Cup at a Time.* New York: Hyperion, 1999; http:/money.cnn.com/magazines/fortune/bestcompanies/2010/snapshots/93.html
35. Ibid., p. 135.
36. Michelli makes much of the issue of predictability at Starbucks. See Joseph A. Michelli. *The Starbucks Experience: 5 Principles for Turning Ordinary Into Extraordinary.* New York: McGraw-Hill, 2007, pp. 99–103.
37. Although Starbucks has been building more variation into the shops; see John Simmons. *My Sister's a Barista: How They Made Starbucks a Home Away From Home.* London: Cyon Books, 2005, p. 112.
38. www.cbsnews.com/stories/2006/04/21/60minutes/main1532246.shtml
39. At least that is the mark-up reported for Dunkin' Donuts coffee; see Stephen Rodrick. "New York Is Suddenly Brimming With Dunkin' Donuts Stores. And With a Starbucks on Every Corner, a Coffee Class War Is Brewing." *New York Magazine,* November 28, 2005. Ironically, Rodrick accused Dunkin' Donuts of "the McDonaldization of coffee" (speed, efficiency, use of nonhuman technology) but misses the fact that Starbucks has done much the same thing, although perhaps not as blatantly as Dunkin' Donuts.
40. "United States of Starbucks." Global News Wire, December 1, 2005.
41. Howard Schultz. *Pour Your Heart Into It: How Starbucks Built a Company One Cup at a Time.* New York: Hyperion, 1999, p. 172.
42. John Simmons. *My Sister's a Barista: How They Made Starbucks a Home Away From Home.* London: Cyon Books, 2005, p. 85.

43. Monica Soto Ouichi. "Opportunity Brewing for Starbucks in China; Starbucks in China—Q & A With Chairman Howard Schultz." *Seattle Times*, October 9, 2005, pp. D1ff; www .btimes.com/ny/Current_News/BTIMES/articles/20090114020036/Article/

44. Peter Enav. "Taiwan Tea Culture's Vitality Slipping Away." *China Post*, September 19, 2006, p. 19.

45. Edward Iwata. "Owner of Small Coffee Shop Takes on Java Titan Starbucks: Lawsuit Could Clarify Cloudy Antitrust Issues." *USA TODAY*, December 20, 2006, pp. 1Bff.

46. Allison Linn. "In Starbucks' Shadow, Smaller Coffeehouses Thrive." Associated Press Financial Wire, September 8, 2006.

47. Howard Schultz. *Pour Your Heart Into It: How Starbucks Built a Company One Cup at a Time.* New York: Hyperion, 1999, p. 208.

48. Janet Adamy. "Getting the Kids Hooked on Starbucks." Associated Press Financial Wire, June 27, 2006.

49. Benjamin Barber. *Consumed: How Markets Corrupt Children, Infantilize Adults, and Swallow Citizens Whole.* New York: W. W. Norton, 2007.

50. Elif Izberk-Bilgin and Aaron Ahuvia. "Ebayization." In George Ritzer, ed. *McDonaldization: The Reader,* 3rd ed. Thousand Oaks, CA: Sage, 2010.

51. George Ritzer. *Enchanting a Disenchanted World: Continuity and Change in the Cathedrals of Consumption,* 3rd ed. Thousand Oaks, Sage, 2010.

52. Elif Izberk-Bilgin and Aaron Ahuvia. "Ebayization." In George Ritzer, ed. *McDonaldization: The Reader,* 3rd ed. Thousand Oaks, CA: Sage, 2010, p. 429.

53. Nathan Jurgenson and George Ritzer. "Efficiency, Effectiveness, and Web 2.0." In Sharon Kleinman, ed. *The Culture of Efficiency: Technology in Everyday Life.* New York: Peter Lang, 2009, pp. 51–67.

Bibliography

R ather than repeating the citations listed in the endnotes, I would like to use this section to cite some of the major (largely) academic works that served as resources for this book. There are three categories of such resources. The first is the work of Max Weber, especially that dealing with rationalization. The second is the work of various neo-Weberians who have modified and expanded on Weber's original ideas. Finally, there is a series of works that focus on specific aspects of our McDonaldizing society.

Works by Max Weber

Economy and Society: An Outline of Interpretive Sociology, edited by Guenther Roth and Claus Wittich, translated by Ephraim Fischoff et al. Berkeley: University of California Press, 1978.

General Economic History, translated by Frank H. Knight. Mineola, NY: Dover, 1927/2003.

The Protestant Ethic and the Spirit of Capitalism, new introduction and translation by Stephen Kalberg, 3rd Roxbury ed. Los Angeles, CA: Roxbury, 2002.

The Rational and Social Foundations of Music. Carbondale: Southern Illinois University Press, 1921/1958.

The Religion of China: Confucianism and Taoism. New York: Macmillan, 1916/1964.

The Religion of India: The Sociology of Hinduism and Buddhism. Glencoe, IL: Free Press, 1916–1917/1958.

"Religious Rejections of the World and Their Directions." In H. H. Gerth and C. W. Mills, eds., *From Max Weber: Essays in Sociology.* New York: Oxford University Press, 1915/1958, pp. 323–359.

"The Social Psychology of the World Religions." In H. H. Gerth and C. W. Mills, eds., *From Max Weber: Essays in Sociology.* New York: Oxford University Press, 1915/1958, pp. 267–301.

Works by Neo-Weberians

Rogers Brubaker. *The Limits of Rationality: An Essay on the Social and Moral Thought of Max Weber.* London: Allen & Unwin, 1984.

Randall Collins. *Weberian Sociological Theory*. Cambridge, UK: Cambridge University Press, 1985.

Randall Collins. "Weber's Last Theory of Capitalism: A Systematization." *American Sociological Review* 45 (1980): 925–942.

Arnold Eisen. "The Meanings and Confusions of Weberian 'Rationality.'" *British Journal of Sociology* 29 (1978): 57–70.

Harvey Greisman. "Disenchantment of the World." *British Journal of Sociology* 27 (1976): 497–506.

Harvey Greisman and George Ritzer. "Max Weber, Critical Theory and the Administered World." *Qualitative Sociology* 4 (1981): 34–55.

Jurgen Habermas. *The Theory of Communicative Action*. Vol. 1, *Reason and the Rationalization of Society*. Boston: Beacon, 1984.

Stephen Kalberg. "Max Weber." In George Ritzer, ed., *The Blackwell Companion to Major Social Theorists*. Oxford, UK: Blackwell, 2000, pp. 144–204.

Stephen Kalberg. *Max Weber's Comparative Historical Sociology*. Chicago: University of Chicago Press, 1994.

Stephen Kalberg. "Max Weber's Types of Rationality: Cornerstones for the Analysis of Rationalization Processes in History." *American Journal of Sociology* 85 (1980): 1145–1179.

Stephen Kalberg. "The Rationalization of Action in Max Weber's Sociology of Religion." *Sociological Theory* 8 (1990): 58–84.

Donald Levine. "Rationality and Freedom: Weber and Beyond." *Sociological Inquiry* 51 (1981): 5–25.

Arthur Mitzman. *The Iron Cage: An Historical Interpretation of Max Weber*, with a new introduction by the author, preface by Lewis A. Coser. New Brunswick, NJ: Transaction Books, 1985.

Wolfgang Mommsen. *The Age of Bureaucracy*. New York: Harper & Row, 1974.

George Ritzer. "Professionalization, Bureaucratization and Rationalization: The Views of Max Weber." *Social Forces* 53 (1975): 627–634.

George Ritzer and Terri LeMoyne. "Hyperrationality." In George Ritzer, ed., *Meta-theorizing in Sociology*. Lexington, MA: Lexington Books, 1991, pp. 93–115.

George Ritzer and David Walczak. "Rationalization and the Deprofessionalization of Physicians." *Social Forces* 67 (1988): 1–22.

Guenther Roth and Reinhard Bendix, eds. *Scholarship and Partisanship: Essays on Max Weber*. Berkeley: University of California Press, 1971.

Lawrence Scaff. *Fleeing the Iron Cage: Culture, Politics, and Modernity in the Thought of Max Weber*. Berkeley: University of California Press, 1989.

Wolfgang Schluchter. *The Rise of Western Rationalism: Max Weber's Developmental History*, translated, with an introduction, by Guenther Roth. Berkeley: University of California Press, 1981.

Mark A. Schneider. *Culture and Enchantment*. Chicago: University of Chicago Press, 1993.

Alan Sica. *Weber, Irrationality and Social Order*. Berkeley: University of California Press, 1988.

Ronald Takaki. *Iron Cages: Race and Culture in 19th-Century America*, rev. ed. New York: Oxford University Press, 2000.

Works on Various Aspects of a McDonaldizing Society

Mark Alfino, John Caputo, and Robin Wynyard, eds. *McDonaldization Revisited*. Westport, CT: Greenwood, 1998.

Benjamin Barber. *Consumed: How Markets Corrupt Children, Infantilize Adults, and Swallow Citizens Whole*. New York: W. W. Norton, 2007.

Benjamin Barber. *Jihad vs. McWorld*. New York: Times Books, 1995.

Zygmunt Bauman. *Modernity and the Holocaust*. Ithaca, NY: Cornell University Press, 2000.

Daniel Bell. *The Coming of Post-Industrial Society: A Venture in Social Forecasting*, special anniversary edition, with a new foreword by the author. New York: Basic Books, 1999.

Max Boas and Steve Chain. *Big Mac: The Unauthorized Story of McDonald's*. New York: E. P. Dutton, 1976.

Daniel J. Boorstin. *The Image: A Guide to Pseudo-Events in America*, with a new foreword by the author and an afterword by George F. Will, 25th anniversary edition. New York: Atheneum, 1987.

Pierre Bourdieu. *Distinction: A Social Critique of the Judgment of Taste*. Cambridge, MA: Harvard University Press, 1984.

Alan Bryman. *Disney and His Worlds*. London: Routledge, 1995.

Alan Bryman. "The Disneyization of Society." *Sociological Review* 47 (1999): 25–47.

Alan Bryman. *The Disneyization of Society*. London: Sage, 2004.

Deborah Cameron. *Good to Talk? Living in a Communication Culture*. London: Sage, 2000.

Simon Clarke. "The Crisis of Fordism or the Crisis of Social Democracy?" *Telos* 83 (1990): 71–98.

Ben Cohen, Jerry Greenfield, and Meredith Mann. *Ben & Jerry's Double-Dip: How to Run a Values-Led Business and Make Money, Too*. New York: Fireside, 1998.

Stanley Cohen and Laurie Taylor. *Escape Attempts: The Theory and Practice of Resistance to Everyday Life*, 2nd ed. London: Routledge, 1992.

Greg Critser. *Fat Land*. Boston: Houghton Mifflin, 2004.

Thomas S. Dicke. *Franchising in America: The Development of a Business Method, 1840–1980*. Chapel Hill: University of North Carolina Press, 1992.

John Drane. *After McDonaldization: Mission, Ministry, and Christian Discipleship in an Age of Uncertainty*. Grand Rapids, MI: Baker Academic, 2008.

John Drane. *The McDonaldization of the Church*. London: Darton, Longman, and Todd, 2001 (a 2008 edition was published by Smyth & Helwys, Macon, GA).

Donna Dustin. *The McDonaldization of Social Work*. Farnham, Surrey, UK: Ashgate, 2008.

Richard Edwards. *Contested Terrain: The Transformation of the Workplace in the Twentieth Century*. New York: Basic Books, 1979.

Morten G. Ender. *American Soldiers in Iraq: McSoldiers or Innovative Professionals?* New York: Routledge, 2009.

Charles Fishman. *The Wal-Mart Effect: How the World's Most Powerful Company Really Works—And How It's Transforming the American Economy*. New York: Penguin, 2006.

Marshall Fishwick, ed. *Ronald Revisited: The World of Ronald McDonald*. Bowling Green, OH: Bowling Green University Press, 1983.

Stephen M. Fjellman. *Vinyl Leaves: Walt Disney World and America*. Boulder, CO: Westview, 1992.

James T. Flink. *The Automobile Age*. Cambridge, MA: MIT Press, 1988.

Henry Ford. *My Life and Work*. Garden City, NY: Doubleday, Page, 1922.

Thomas Friedman. *The World Is Flat: A Brief History of the 21st Century*. New York: Farrar, Strauss, Giroux, 2005.

Thomas L. Friedman. *The Lexus and the Olive Tree*, rev. ed. New York: Farrar, Straus, Giroux, 2000.

Herbert J. Gans. *The Levittowners: Ways of Life and Politics in a New Suburban Community*, with a new preface by the author. New York: Columbia University Press, 1967/1982.

Barbara Garson. *All the Livelong Day: The Meaning and Demeaning of Routine Work*, rev. and updated ed. New York: Penguin, 1994.

Steven L. Goldman, Roger N. Nagel, and Kenneth Preiss. *Agile Competitors and Virtual Organizations: Strategies for Enriching the Customer*. New York: Van Nostrand Reinhold, 1995.

Richard E. Gordon, Katharine K. Gordon, and Max Gunther. *The Split-Level Trap*. New York: Gilbert Geis, 1960.

Roger Gosden. *Designing Babies: The Brave New World of Reproductive Technology*. New York: W. H. Freeman, 1999.

Harold Gracey. "Learning the Student Role: Kindergarten as Academic Boot Camp." In Dennis Wrong and Harold Gracey, eds., *Readings in Introductory Sociology*. New York: Macmillan, 1967.

Allen Guttmann. *From Ritual to Record: The Nature of Modern Sports*. New York: Cambridge University Press, 1978.

Jeffrey Hadden and Charles E. Swann. *Prime Time Preachers: The Rising Power of Televangelism*. Reading, MA: Addison-Wesley, 1981.

Jerald Hage and Charles H. Powers. *Post-Industrial Lives: Roles and Relationships in the 21st Century*. Newbury Park, CA: Sage, 1992.

David Harvey. *The Condition of Postmodernity: An Enquiry Into the Origins of Cultural Change*. Oxford, UK: Basil Blackwell, 1989.

Dennis Hayes and Robin Wynyard, eds. *The McDonaldization of Higher Education*. Westport, CT: Bergin & Garvey, 2002.

Elif Izberk-Bilgin and Aaron Ahuvia. "Ebayization." In George Ritzer, ed. *McDonaldization: The Reader*, 3rd ed. Thousand Oaks, CA: Sage, 2010.

Kathleen Jamieson. *Eloquence in an Electronic Age: The Transformation of Political Speechmaking*. New York: Oxford University Press, 1988.

Robert Kanigel. *One Best Way: Frederick Winslow Taylor and the Enigma of Efficiency*. New York: Viking, 1997.

Joe L. Kincheloe. *The Sign of the Burger: McDonald's and the Culture of Power*. Philadelphia: Temple University Press, 2002.

Aliza Kolker and B. Meredith Burke. *Prenatal Testing: A Sociological Perspective*. Westport, CT: Bergin & Garvey, 1994.

William Severini Kowinski. *The Malling of America: An Inside Look at the Great Consumer Paradise*. New York: William Morrow, 1985.

Jon Krakauer. *Into Thin Air.* New York: Anchor, 1997.

Ray Kroc. *Grinding It Out.* New York: Berkeley Medallion Books, 1977.

Corby Kummer. *The Pleasures of Slow Food: Celebrating Authentic Traditions, Flavors, and Recipes.* San Francisco: Chronicle Books, 2002.

Raymond Kurzweil. *The Age of Intelligent Machines.* Cambridge, MA: MIT Press, 1990.

Fred "Chico" Lager. *Ben & Jerry's: The Inside Scoop.* New York: Crown, 1994.

Frank Lechner and John Boli, eds. *The Globalization Reader,* 2nd ed. Oxford, UK: Blackwell, 2004.

Robin Leidner. *Fast Food, Fast Talk: Service Work and the Routinization of Everyday Life.* Berkeley: University of California Press, 1993.

John F. Love. *McDonald's: Behind the Arches,* rev. ed. New York: Bantam Books, 1995.

Stan Luxenberg. *Roadside Empires: How the Chains Franchised America.* New York: Viking, 1985.

Jean-François Lyotard. *The Postmodern Condition: A Report on Knowledge.* Minneapolis: University of Minnesota Press, 1984.

Frank Mankiewicz and Joel Swerdlow. *Remote Control: Television and the Manipulation of American Life.* New York: Time Books, 1978.

Joseph A. Micheli. *The Starbucks Experience: 5 Principles for Turning Ordinary Into Extraordinary.* New York: McGraw-Hill, 2007.

Jessica Mitford. *The American Way of Birth.* New York: Plume, 1993.

Ian I. Mitroff and Warren Bennis. *The Unreality Industry: The Deliberate Manufacturing of Falsehood and What It Is Doing to Our Lives.* New York: Oxford University Press, 1993.

Jerry Newman. *My Secret Life on the McJob: Lessons From Behind the Counter Guaranteed to Supersize Any Management Style.* New York: McGraw-Hill, 2007.

Sherwin B. Nuland. *How We Die: Reflections on Life's Final Chapter.* New York: Knopf, 1994.

Lauren L. O'Toole. "McDonald's at the Gym? A Tale of Two Curves." *Qualitative Sociology* 32 (2009): 75–91.

Martin Parker and David Jary. "The McUniversity: Organization, Management and Academic Subjectivity." *Organization* 2 (1995): 319–337.

Stacy Perman. *In-N-Out Burger.* New York: Collins Business, 2009.

Thomas J. Peters and Robert H. Waterman. *In Search of Excellence: Lessons From America's Best-Run Companies.* New York: Harper & Row, 1982.

Neil Postman. *Amusing Ourselves to Death: Public Discourse in the Age of Show Business.* New York: Viking, 1985.

Neil Postman. *Technopoly: The Surrender of Culture to Technology.* New York: Knopf, 1992.

Peter Prichard. *The Making of McPaper: The Inside Story of* USA TODAY. Kansas City, MO: Andrews, McMeel and Parker, 1987.

Stanley Joel Reiser. *Medicine and the Reign of Technology.* Cambridge, UK: Cambridge University Press, 1978.

Ester Reiter. *Making Fast Food: From the Frying Pan Into the Fryer,* 2nd ed. Montreal: McGill-Queen's University Press, 1997.

George Ritzer. "The McDonaldization of Society." *Journal of American Culture* 6 (1983): 100–107.

George Ritzer. *Expressing America: A Critique of the Global Credit Card Society.* Newbury Park, CA: Sage, 1995.

George Ritzer. *The McDonaldization Thesis.* London: Sage, 1998.

George Ritzer, ed. "McDonaldization: Chicago, America, the World" (Special issue). *American Behavioral Scientist* 47 (October 2003).

George Ritzer. *The Globalization of Nothing 2.* Thousand Oaks, CA: Sage, 2007.

George Ritzer. *Enchanting a Disenchanted World: Revolutionizing the Means of Consumption,* 3rd ed. Thousand Oaks, CA: Sage, 2010.

George Ritzer, ed. *McDonaldization: The Reader 3.* Thousand Oaks, CA: Sage, 2010.

George Ritzer, Paul Dean, and Nathan Jurgenson. "The Coming of Age of the Prosumer" (Special issue). *American Behavioral Scientist,* forthcoming, 2012.

George Ritzer and Nathan Jurgenson. "Production, Consumption, Prosumption: The Nature of Capitalism in the Age of the Digital 'Prosumer.'" *Journal of Consumer Culture* 10, no. 1 (2010): 13–36.

George Ritzer and David Walczak. "The Changing Nature of American Medicine." *Journal of American Culture* 9 (1987): 43–51.

Roland Robertson. *Globalization: Social Theory and Global Culture.* London: Sage, 1992.

Chris Rojek. *Ways of Escape: Modern Transformations in Leisure and Travel.* London: Routledge, 1993.

Eric Schlosser. *Chew on This: Everything You Don't Want to Know About Fast Food.* Boston: Houghton Mifflin, 2007.

Eric Schlosser. *Fast Food Nation.* Boston: Houghton Mifflin, 2001.

Howard Schulz. *Pour Your Heart Into It: How Starbucks Built a Company One Cup at a Time.* New York: Hyperion, 1997.

Charles E. Silberman. *Crisis in the Classroom: The Remaking of American Education.* New York: Random House, 1970.

John Simmons. *My Sister's a Barista: How They Made Starbucks a Home Away From Home.* London: Cyan Books, 2005.

Bryant Simon. *Everything but the Coffee: Learning About America From Starbucks.* Berkeley: University of California Press, 2009.

Peter Singer. *Animal Liberation,* 2nd ed. New York: New York Review of Books, 1990.

Alfred P. Sloan Jr. *My Years at General Motors.* Garden City, NY: Doubleday, 1964.

Barry Smart, ed. *Resisting McDonaldization.* London: Sage, 1999.

Morgan Spurlock. *Don't Eat This Book: Fast Food and the Supersizing of America.* New York: G. P. Putnam, 2005.

Frederick W. Taylor. *The Principles of Scientific Management.* New York: Harper & Row, 1947.

John Vidal. *McLibel: Burger Culture on Trial.* New York: New Press, 1997.

James L. Watson, ed. *Golden Arches East: McDonald's in East Asia.* Stanford, CA: Stanford University Press, 1997.

Shoshana Zuboff. *In the Age of the Smart Machine: The Future of Work and Power.* New York: Basic Books, 1988.

Index

Nirula's (India), 4
Nonhuman technology:
  airline pilots, 110
  assembly line production, 37, 108
  customer service
      representatives, 109
  defined, 15, 102
  disenchantment, 129
  education, 106
  employee control, 103–110
  fast-food industry, 103–105
  health care, 106–107
  Holocaust example, 34
  of Holocaust, 34
  scientific management, 34, 37
  supermarkets, 109
  telemarketing, 109–110
  See also Dehumanization
Norman, Al, 150, 151
NutriSystem, 59

Obama, Barack, 83
Office Depot, 4
Ohio, 40
Old Navy, 97
Olive Garden, 3, 75
One-best-way, 35
Outback Steakhouse, 3

Papa John's, 90
Pearle Vision, 3, 69
PetSmart, 3
Philippines, 5
Pinto (Ford Motor Company), 86
Pittsburgh Steelers, 81
Pizza Hut:
  customer tasks, 70
  international locations, 4, 7
  McDonaldization of, 2–3, 14
Pleasantville (1998), 91–92
Poland, 152
Politics:
  calculability, 83
  irrationality of rationality, 127
  quantification, 83
Pollo Campero, 5
Pollo Tropical, 5
Poltergeist (1982), 91
Post-Fordism, 48–49

Postindustrialism, 47–48
Postmodernism, 49–50, 195n67
Predictability:
  amusement parks, 96, 101, 201n42
  assembly line production, 36–37
  camping, 101–102
  characteristics of, 14–15
  disenchantment, 129
  eBay, 180
  employee behavior, 94–96
  employee scripts, 92–94
  family fun centers, 100–101, 102
  fast-food industry, 14, 87, 89–90,
      92–93, 94, 95–96, 97–98
  film industry, 91–92, 98
  haircutting salons, 92
  IKEA, 18
  In-N-Out Burger, 22, 90
  McDonald's, 14, 87, 89–90, 92–93,
      94, 95–96, 97–98
  motel chains, 88–89
  of bureaucratization, 31
  products and processes, 97–100
  religious service, 90
  safety concerns, 100–102
  scientific management, 31, 36
  shopping malls, 100
  sports, 100
  suburban housing, 90–92
  telemarketing, 94
  television, 98–99
  travel industry, 99
  Web 1.0 technology, 182
  workplace bureaucracy, 90
Pregnancy, 117–119
Pret A Manger:
  calculability, 26
  control, 25–26
  efficiency, 25
  irrationality of rationality, 26
  McDonaldization of, 5, 25–26
  non-McDonaldized operations,
      25, 26
  predictability, 26
Prospective payment, 80, 106
Prosumers, 69
Psycho (1960), 88, 98
Publishing industry, 62
Puerto Rico, 5

About the Author

George Ritzer is Distinguished University Professor at the University of Maryland, where he has also been a Distinguished Scholar-Teacher and won a Teaching Excellence Award. He was awarded the 2000 Distinguished Contributions to Teaching Award by the American Sociological Association, and he has been named the Robin M. Williams Lecturer by the Eastern Sociological Society for 2012–2013. He has an honorary doctorate from LaTrobe University in Australia, and in 2012 he will be accorded the same honor by Oxford Brookes University. He is best known for *The McDonaldization of Society* (translated into more than a dozen languages) and *McDonaldization: The Reader 3* as well as several related books, including *Expressing America: A Critique of the Global Credit Card Society, Enchanting a Disenchanted World 3, The Globalization of Nothing 2,* and *Globalization: A Basic Text*. He is the Editor of the *Encyclopedia of Social Theory* (2 vols.) and the *Encyclopedia of Sociology* (11 vols.), is currently editing the *Encyclopedia of Globalization* (5 vols.), and is Founding Editor of the *Journal of Consumer Culture*.

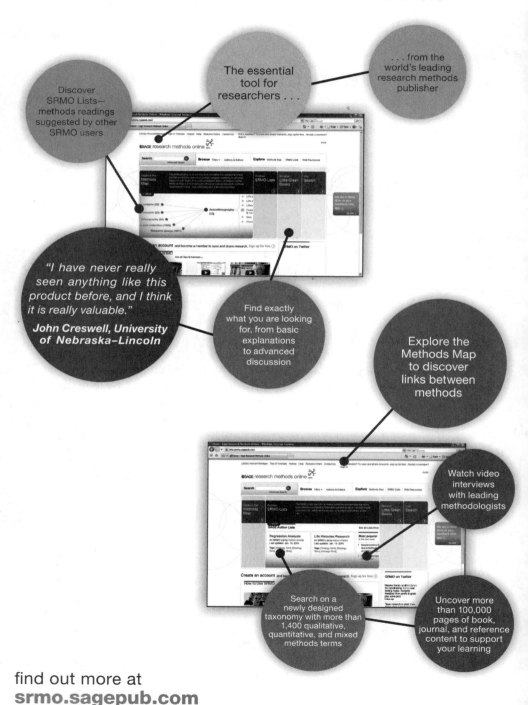

ⓈSAGE research**methods**
The Essential Online Tool for Researchers

Discover SRMO Lists— methods readings suggested by other SRMO users

The essential tool for researchers . . .

. . . from the world's leading research methods publisher

"I have never really seen anything like this product before, and I think it is really valuable."
John Creswell, University of Nebraska–Lincoln

Find exactly what you are looking for, from basic explanations to advanced discussion

Explore the Methods Map to discover links between methods

Watch video interviews with leading methodologists

Search on a newly designed taxonomy with more than 1,400 qualitative, quantitative, and mixed methods terms

Uncover more than 100,000 pages of book, journal, and reference content to support your learning

find out more at
srmo.sagepub.com